ONLY A SOLDIER KNOWS

Ray Lane is a former commanding officer of the Irish Defence Force Ordnance School. On his retirement, he had been a member of the Irish Defence Forces for 45 years. He has served overseas in many missions with the UN, EU and NATO in Lebanon, Bosnia, Afghanistan and Somalia. He has served as an expert witness on the International Criminal Tribunal for the Former Yugoslavia (1993), the UN Fact-Finding Mission on the Gaza Conflict (2009) and the Callanan Review (2011). Today he is an advisor for UNOPS Ukraine – tasked with freeing up land from landmines and devising a strategy for the developing improvised explosive device situation. He also works for a start-up company at the University of Galway that is looking at developing natural capabilities in dealing with wound care and other military capability deficiencies in the field of chemical and biological agents. He lives in Naas, Co. Kildare.

ONLY A SOLDIER KNOWS

LIFE ON THE FRONT LINES WITH THE IRISH DEFENCE FORCES

RAY LANE

Gill Books

Gill Books
Hume Avenue
Park West
Dublin 12
www.gillbooks.ie

Gill Books is an imprint of M.H. Gill and Co.

9781804580561

Design origination by O'K Graphic Design, Dublin
Typeset by Typo•glyphix, Burton-on-Trent, DE14 3HE
Edited by Alison Walsh
Copy edited by Esther Ní Dhonnacha
Proofread by Tamsin Shelton
Printed and bound in Great Britain by Clays Ltd, Elcograf S.p.A.
This book is typeset in 12/18 pt Minion.

The paper used in this book comes from the wood pulp of managed forests.
For every tree felled, at least one tree is planted, thereby renewing natural resources.

A CIP catalogue record for this book is available from the British Library.

5 4 3 2 1

To the Officers, NCOs and men and women of the Ordnance Corps who have given this state the most professional service over many decades in some really difficult situations, without the official recognition they most definitely deserve. This book is written in their honour, with my absolute admiration for their service.

CONTENTS

PROLOGUE

Whenever I tell people that I'm a retired bomb-disposal officer, they always mention *The Hurt Locker*. I don't blame them: the images of guys sweating in their bomb suits with only seconds to spare before they get blown up are pretty exciting. The reality is a little bit different. We don't just pitch up to a bomb and go about dismantling it: as a result of intensive training, there's a process involved that is there to keep us and the public safe. And the rules, such as the mandatory 30-minute break between each manual approach to the device, would not make for an exciting action movie. However, bomb disposal is dangerous. Of course it is. If you are dealing with 10 beer kegs filled to the brim with explosives on the border, a fiendish homemade device in Lebanon, or a crude wooden improvised explosive device (IED) in Afghanistan that still might cause many casualties – that's the very definition of hazardous. But if I was ever to think about that, I wouldn't be able to do my job. The process and the training we've undertaken are what keeps me and my colleagues safe.

The first rule of bomb disposal is that you never take unnecessary risks. As you can imagine in a job where a mistake can end in a fatality, it pays to do it properly. That's why I can remember my very first callout as if it were yesterday. It was 1978 and I was only recently qualified as a bomb-disposal officer in the Defence Forces Ordnance School, so I was still pretty much a rookie. This was during the depths of the Troubles, so, as you can imagine, we were busy. It was a gloomy Sunday evening when I was called out to Bawnboy, Co. Cavan, to defuse a 1,000kg bomb, planted by the IRA. It had been hidden in a ditch, ready to be moved to the attack site, and a warning had been phoned in to the Gardaí. In spite of the sheer size of the bomb, I was confident that I knew what I was doing. I'd had my training, both in the Irish and British armies, dealing with hypothetical scenarios from a bomb on a train filled with fuel to a truck filled with mortar bombs, and I was ready for it. Or so I thought.

I hopped in the helicopter (our usual method of transport to distant callouts) that would take me from Finner Camp in Ballyshannon, Co. Donegal, to Bawnboy to see what I was dealing with. I could see the pocket-sized fields and hedgerows of the Border as we flew along, wondering what was down there, hiding in farmyards and small villages. The pilot landed the helicopter on the local GAA pitch, where I was greeted by the Gardaí and taken to the scene. I looked from a distance at the 10 beer kegs full to the brim with homemade explosives, and even though my heart began to race with a mixture of fear and excitement, my training came into play and my brain began to turn over potential scenarios to defuse it.

In our business, we try to do as much as we can from a distance, for obvious reasons. This means sending in a robot, which you control remotely, to look at the device. Here, however, the device was totally inaccessible, so no robot was going to tell me how to tackle it. I was going to have to take a manual approach to it. This is the last resort for anyone in my line of work, so as I donned my bomb suit, I felt my mouth dry with anticipation. Still, as per procedure, I got out my map and I outlined the danger area for the Gardaí, so that they could evacuate it. Given the size of the bomb, it was a huge area, dotted with small farms and houses, and I left them to it while I set to work.

It didn't start well. There were a number of approaches that I could have taken, but I decided that I was going to try to burn the explosive out of the containers. We had just got new disruptive charges for that purpose, so I was confident of success. But as the explosive started to burn in the container and it went further down, there was less oxygen, so guess what? It stopped burning. Stopped dead. The question was what state was it in now? Had I made it worse?

I thought for a good, long while and discussed it with my team. Then I made a decision: I would blow the bomb up *in situ*. I would set up a charge, attach it to the containers and boom, to use the technical word. Thankfully, I thought, the area was deserted, so the damage to any buildings would be minimal. I cleared the area of Gardaí and any army personnel who might be lingering and prepared the charge. Then I was on my own. In bomb disposal, there is always that long walk to the device, alone. It's essential for safety, but it's a part of the job that we never quite get used to. Now, listening to my breath and to the hiss of the oxygen supply in my heavy bomb suit, I walked towards the kegs and knelt down to attach the charge to the bomb. Then, in spite of the human instinct telling me to run, I walked slowly back.

The explosion was huge, sending a great cloud of smoke into the air, making our ears ring with the sheer volume of the blast. Still, I thought, as the dust settled, at least it's done – and, more importantly, no one has been injured.

Then the door of a little cottage a hundred or so metres away opened, very slowly. And out came an elderly man, resting on a Zimmer frame. He looked a bit surprised at having been woken from his nap. I looked at him and my legs turned to jelly.

———◆———

Forty years and many, many callouts later, I came to my very last job. It was 2011 and by this stage, I had seen just about everything, so a pipe bomb planted in a gang-related dispute in Dolphin's Barn, Dublin 8, should have been a piece of cake. I was 58, way beyond the normal retirement age for a bomb-disposal guy, but due to a shortage of qualified officers at the time, I was hanging on in there. My eyesight wasn't as sharp as it had once been, but at this stage, I could defuse pipe bombs in my sleep.

I was on duty in Cathal Brugha Barracks in Rathmines when the call came. In the past, my adrenaline would have been motoring. I'd have been in mighty form. My team and I would get together and formulate a plan for the task ahead, then we'd head to the vehicle. But on this night, I had no energy. I wasn't feeling it. What's more, I'd been to the optician two weeks before and been fitted for a pair of glasses – my failing eyesight would make my job all the more difficult, so I wasn't happy. Even though I'd passed my most recent two-yearly evaluation, the NCT for bomb-disposal officers if you like, I was over the hill and I knew it. In fact, the hill was now receding into the distance.

Nonetheless, there was work to be done, so I briefed the guys on my team, and we took off in our specially adapted van with our Garda motorbike outriders to bring us up the busy Grand

Canal to the location. There, I went through the normal procedure, evacuating the whole road, setting up an incident control point (ICP), discussing all potential outcomes with my team and planning our approach to making the bomb safe. We'd start by getting our remote robot out and sending it off towards the car, while watching its progress closely on the screens in front of us.

Now, whoever had made the pipe bomb clearly knew exactly how far under the car the arm of my HOBO robot would go, because they had placed it just out of reach. However, robots are fitted with cameras, so we got a good picture of it, then I brought the HOBO back. With the pipe bomb inaccessible, we needed to think of a plan B.

The guys loved to take the mick out of me, so one of them said cheekily, 'What are you going to do now, sir?' But another member of the team, whom I trusted completely, looked at me and said, 'You're not with it tonight.'

'No, I'm not,' I replied. 'Keep an eye on me.' I have always applied this rule to my job: if one of the team thinks I'm going to do something wrong, they tell me. They just say, 'Stop.' Bomb disposal requires a clear head and if one of us is off, it can be a disaster. In any case, my training means that I bounce things off my team. Bomb disposal is a real team sport. The training of all team members is of the highest international standard and assessed on a regular basis to ensure that we are always match fit. This is the reason that the Ordnance School has become a recognised centre of excellence with students from all over the world.

Still, I wanted out of there. I wanted it over with. The appointment in the opticians, and the reminder that I was getting on, had rattled me and I knew that I wasn't right. In the end, though, I had to get into my bomb suit and go forward to examine the bomb more closely.

A bomb suit is not comfortable. Apart from the weight of the thing, a full 80lbs, even with a cooling system, it's hot as hell. In order to wear it you need to be fit, agile and, in my opinion, young. The bomb suit is mandatory for bomb-disposal situations unless you are in a Category A situation where there is a real and imminent danger of loss of life and the suit would slow the operation down.

We also carried what's known as a 'CV reacher' in our hand: that's a steel grabber, rather like a litter picker, but it has a spring-loaded 'V', which can be attached to the bomb to pull it out.

I got into my suit, and I slid under the car, but I couldn't see that clearly because I wasn't wearing my brand-new glasses. So I struggled back to my feet and walked slowly back to the van, not happy at all. When you walk back, one of the team comes towards you to open the suit's visor, just to relax you. It's oppressive inside the helmet.

One of my team, Stevie McCabe, lifted the visor and I said angrily, 'Get. My. F*cking. Glasses.'

He laughed and ran to get them, but when he returned, he warned me, 'Now, you're not going back there for a while.' The 'soak time' had kicked in. This mandatory waiting period varies between half an hour and two hours, depending on the device, and allows time for further assessment of its capabilities, the timing device attached, what material is being used and so on. Even though I was impatient to get it all over with, I knew the rules, so we went back to the truck, where I took off the helmet completely and sat down.

'What's with the glasses, sir?' Stevie asked.

'Stevie, I should have gone to Specsavers,' I replied, trying to make a joke out of my earlier visit to the optician. This seemed to amuse everyone except me, but we relaxed a bit over a cup of coffee, before I put my glasses back on, then the helmet, pulling the visor down and going forward. With my glasses, it all fitted into place. I pushed myself as far as I could go underneath the car and then disrupted the pipe bomb with a small charge to prevent it going off, to render it safe. I carefully bagged all the bomb components for forensics, as per procedure, and then I returned to the truck. I took off my helmet and I sat down.

I said to Stevie, 'It's over.'

I was finished. I knew that. The incident had ended happily, but I knew it wasn't my finest hour. I should have paid more attention to the placement of the pipe bomb way back at the incident control point. I should have spent more time assessing it and modifying any remote means to get at it before I took it on manually. Ten years before, I'd have said, 'Give me a sweeping brush.' I'd have taped it to the arm of the robot, extended the arm in and pulled out the device before approaching it manually. When I'd gone forward in my bomb suit, I'd put myself in unnecessary danger.

I decided there and then that after more than forty years in the Irish Army doing explosive ordnance duties, I was never doing another task. Wisely, Stevie didn't say anything then, but later he would tell me that he knew by my behaviour that day that I was done. I got back into the truck, went back to Cathal Brugha Barracks and rang Bridget, my wife, and I said, 'It's over, that's the end of it. I'm not doing any more of this. I'm fed up.' And that was it.

It might seem like an abrupt end to a career, but I knew that I was doing the right thing. If I couldn't do my job properly, it was time to hang up my hat and leave it to someone fresh. I'd only agreed to stay because of training gaps, but I knew that now it was time to go.

What did I learn in my long career in the Irish Defence Forces, I wondered as I trudged wearily back to my office? What was it all for? I suspect that we all ask ourselves this question, no matter what kind of job we do. Recently, I went to Ukraine with UNOPS (United Nations Office for Project Services) to consult on helping them to tackle their huge minefield and developing IED problem. It occurred to me that I'd been here before, many times. From Northern Ireland in the 1970s to Lebanon, to Bosnia and then on to Afghanistan and, briefly, Somalia, I'd seen the same violence and the same despair. I'd seen the same mistakes made and the same cover-ups of those mistakes, instead of learning from them. At times, I'd wonder if we've ever learned anything as a human race. Who was it who said, 'Those who do not learn history are doomed to repeat it'?

It would be easy, in spite of all the high points of my career, to despair. But all I could do, can do, is represent my country faithfully. I've been part of teams eager to protect and to serve, to keep local people safe and to do what we can to help them. I've worked with fantastic people who have given their whole lives to making things better for others. To making a difference. War and conflict are outside my control, but I can help people in a practical way, and every time I kneel down to examine a device or negotiate a ceasefire between hostile forces, I remind myself of this.

CHAPTER 1

NEVER A YES MAN

n 1979, I was a newly qualified officer with the Defence Forces of Ireland Ordnance Corps, when we were ushered into the conference room in Clancy Barracks for a briefing. Pope John Paul II was to visit Ireland on 29 September, and, we were told, this was one of the biggest occasions in the history of the State. At the time, it could be said that the Defence Forces in Ireland was a very Catholic organisation, so you can imagine the excitement. This was our chance to show the Defence Forces at its very best, undertaking bomb-disposal duties in the Phoenix Park, Drogheda, Galway and so on.

The colonel gave us a pep talk about representing the institution properly and then proceeded to allocate personnel to locations. The very first name on the list was mine. At the words 'Lieutenant Ray Lane', there was a snigger in the room. The colonel looked up, then said to me, 'Is that okay?'

'Yes, sir, of course,' I replied. There was another round of sniggering. I was known to be a little anti-Church, so it gave my colleagues a good laugh when they heard that yours truly would be sent to the Phoenix Park to join the million Irish people estimated to be attending the Mass. I was given my orders, to be responsible for any device found in the Park, and I would obey those orders.

Now there's a frequent misunderstanding about the military: that you do what you're told, or else, but that's not actually the case. I can still remember an early lesson from my training, when a colonel said to the 49th cadet class, 'You never do anything unless you know the reason why.' I've stuck by that ever since. You can question an order, get clarity on it and understand it, as I have done many times. Having said that, if you're given an order that you don't like, but that isn't illegal, you can't refuse to do it. That's military law.

So, me and my brand-new robot would be heading to the Phoenix Park. Until then, we'd been using British-made robots, which weren't as sophisticated, but we'd developed one of our own thanks to Winn Technologies, and it had arrived just one week before the Pope. Into Clancy Barracks it came, a bright yellow machine on six wheels, bristling with cameras, with a claw to grab the device and a shotgun to disable the explosive charge. The claw was our pride and joy, because it moved like a human hand and had more or less the same dexterity. It could rotate, open and close, grab and so on. It even had a brand-new Ford Transit van to go with it, with a fancy door that dropped down!

We had been using a British machine called the wheelbarrow, which didn't have a claw and was pretty rudimentary, so to be up to speed with, in fact to overtake, the British Army was something we quietly enjoyed. And it would be perfect for reaching into bins and looking around them through the claw's camera from my perch in the van. The Gardaí had done a risk assessment of the Park and they'd pointed out the bins as obvious objects of concern and other vulnerable areas. Perhaps this mightn't be the first thing that springs to mind during a Papal visit, but you never knew who might use the moment to try something. Only two years later, that would actually happen, when Mehmet Ali Ağca attempted to assassinate the Pope in St Peter's Square.

I can still remember getting to the Park at three o'clock in the morning and seeing the thousands of people outside waiting to be allowed in. The entrance gates had had to be widened; such was the expected volume of people. We drove up to our allocated spot, which was near the Papal Mass site, and set ourselves up. At three-thirty, we got the HOBO out and drove it around the location where the Pope would say Mass. The little red light was on, blinking in the darkness, as we moved it around, nosing the claw into bins and bushes. Somewhere out in the crowds, my mother was sitting on her deckchair, in the colours of the Papal flag, waiting for what must have been one of the biggest moments of her life. And her son would be one of the security staff protecting the Pope. I knew that I'd be making her proud.

I had never seen so many cardinals in my life as there were gathered in three large tents next to the stage. As my robot did the rounds, they leaned over in their full regalia, all curiosity, asking me what it did and how it all worked. And while I mightn't have been the biggest fan of the Church, I couldn't help but admire the effort that had gone into the occasion from both religious and lay organisations.

When the sun rose, the sheer volume of people was a remarkable sight. It was like the biggest outdoor rock concert I'd ever seen, crowds stretching towards the horizon. The next thing, this Aer Lingus jumbo jet flies over us, accompanied by Air Corps Foga CM.170 Magister jets on the tip of each wing, and then the army No. 1 Band started playing 'He's Got the Whole World in His

Hands'. I remember saying to my sergeant that day, 'If they release that as a single now, it'll go to number one in the charts'. Really. Everybody started singing it, all one million people.

Finally, after the jet had touched down at Dublin Airport and the Pope had kissed the soil of Ireland, he arrived in his Popemobile, working his way through the vast crowds to the stage to say Mass, and suddenly, it was a whole other league to a rock concert. It was something quite remarkable, I thought as I wandered around, getting a complete sense of what was going on, waiting for any taskings from the Gardaí. When they'd tell me about an object in a bin, I would hop in my van and set the robot off towards the object, watching it on my monitor until it reached the bin. Then I was able to bring the robot forward, push the arm out and grab the object with the claw, taking a good look at it through the camera. It wasn't foolproof, being brand-new technology, but having that arm represented a little bit of divine intervention. The alternative, which would have been to go forward in my bomb suit and rummage around in a bin, wouldn't have been a good look during the Papal Mass.

The Pope said his famous Mass and then left to go to Drogheda to make that unforgettable speech: 'On my knees I beg you to turn away from the paths of violence and to return to the ways of peace. You may claim to seek justice. I too believe in justice and seek justice. But violence only delays the day of justice. Violence destroys the work of justice. Further violence in Ireland will only drag down to ruin the land you claim to love and the values you claim to cherish'. Hearing that later, and after I had worked for many years along the Border, I knew exactly what he was talking about.

I might not have been the most ardent Catholic, but I was proud to be representing the military there that particular day. And I could see how much the occasion had meant to my mother. When I went home after my couple of days on duty, she was full of questions. Did I get close to him, had I seen him, etc.? I knew that it was one of the highlights of her life. And the event had been totally peaceful. The people arrived in an orderly fashion; the people left in an orderly fashion. And the number of people who came over to me to thank me for being on duty and asked me about the HOBO was both surprising and gratifying. I can still remember bringing the robot out to the side of the vehicle, so everybody could have a look. 'I'm proud to say that it's an Irish robot, built and designed in Ireland, the first of many.' The HOBO had been built to our specifications by Winn Technologies in Kilbrittain, Co. Cork, proving that we Irish could produce world-class equipment in this field. (This work continues to this day. In fact, we have two world-class companies developing humanitarian robots, or unmanned ground vehicles as we call them: ICP NewTech and Reamda. Both are at the cutting edge of this technology and need active government/agency support to introduce their capabilities to the global market. This has nothing to do with neutrality – it is solely about developing superb capability and creating high-spec jobs for Irish people.)

I remained on duty in the Phoenix Park until the Pope returned that night to the Papal Nuncio's residence, when I retired to the less-salubrious setting of Coolock Garda Station, where I was to stay that night. And my very first tour of duty was over.

———◆———

I mention my mother here because without her, I would possibly never have joined the Defence Forces. My mother was always my cheerleader and my supporter and without her, I doubt I would have achieved as much in life. I was a bright kid, but unfocused. I think I might have ended up fighting my way through life somehow, but I would have been unhappy. She understood that all that energy I had needed channelling. From my very first days in primary school, I wasn't content to just sit there and let the world drift by. I wanted to know things; I wanted answers. I wanted to satisfy my curiosity about why things worked the way they did. I had a restlessness, an energy that was hard to contain in primary school in the 1950s and 1960s. So, I was permanently in trouble, and school became a battle between me and the teacher: one which he was determined to win. I can still remember him asking me, every single day, which of the six canes in his cupboard I would like him to use on me first.

Of course, I did try, because what child wants to get beaten every day, but I rarely succeeded. I'd be able to keep quiet and work away on my copybooks for a couple of hours, but before too long, I'd be off again, putting my hand up and asking awkward questions. One incident seems to sum up my time in primary school. We were to make our Confirmation and in honour of the occasion, we were to receive a visit from the Archbishop of Dublin, John Charles McQuaid. For those of you who don't know him, he was a friend of Éamon de Valera, a holder of huge influence at the time, and had played a key role in the drafting of the Irish Constitution, as well as the expansion of the Catholic Church. Receiving a visit from him was a huge honour and we were all to be on our best behaviour.

I should mention here that my parents had a mixed marriage, which was unusual at the time. My father was a Presbyterian from Belfast, Mammy a Catholic from Cork. They'd met while he was installing lighting in Cork's central postal office and even though he had had to convert to Catholicism, he remained true to his own beliefs. As a family, we would alternate between Mass and Sunday service in the nearby Presbyterian church in Rathfarnham, whose cheerful hymns and English-language chats I much preferred to the stuffy Catholic Mass, which was still in Latin at the time.

Picture the scene as the black-clad bishop was ushered into the classroom, to be faced with rows of beaming children. The well-behaved pupils had been placed at the front and I skulked at the back with the bold boys, well out of view. However, to our surprise, McQuaid made a beeline for the back of the class. Sweeping his eyes over us, his gaze alighted on yours truly. He pointed a bony finger at me. 'Give me the name of a hymn with the word "saint" in it,' he said.

'"When the Saints Come Marching In",' I blurted in response, quoting the title of my favourite of the hymns we'd sing on Sundays at Daddy's church.

You could have heard a pin drop. All the colour drained from my teacher's face. McQuaid glared at me, before turning to my teacher. 'Is this the level of Catholic teaching in this school?' Then he swept out of the classroom.

The beating I got that day was particularly vicious. When I got home from school, I made sure to hide my bruised hands, because I knew that Mammy would be upset. She'd been through all of my ups and downs in primary school and felt the injustice of my treatment keenly.

Later that evening, the family gathered around the dinner table as we always did. Daddy asked my three siblings about their days, then turned to me. I muttered something about McQuaid's visit and having to pick a hymn with the word 'saint' in it.

'Which hymn did you pick?' Daddy asked.

'"When the Saints Come Marching In", I mumbled.

Daddy looked at Mammy, then exploded in laughter. I had no idea what he was laughing about, but he seemed to find my choice hilarious. He kept repeating the name of the hymn over and over and guffawing with laughter. Then he said, 'Why are you hiding your hands?'

Reluctantly, I put them on the table. Daddy stared at the cuts and bruises and said nothing. The following day, he accompanied me to school, which was unheard of at the time, and marched into the school office. I waited outside, but I could hear every word of his tirade against the school and the teacher. I was left alone for a bit after that.

Even now, more than sixty years later, I can still remember going into that classroom every day as if I was going to jail, waiting for the teacher to call me out and punish me. Mammy and my brother and sisters suffered almost as much as I did. My sister Anne often tells me how rough I was with her and my brothers Philip and Fred as a child, and I think it must have been the frustration coming out. No child knows how to be compliant naturally, but I can now see that forcing this compliance on us was counterproductive. My only sin as a seven-year-old was to be a live wire and beating it out of me was a misguided strategy. Not so long ago, I went to my grandchild Bobby's school to talk to them about my experiences, and I couldn't believe the difference in schooling. It was so positive and friendly. I was astonished.

After that, secondary school was a revelation. On my first day at De La Salle secondary school in Churchtown, the French teacher swept in, wearing the regulation black gown over his clothes. 'Now,' he said cheerfully, 'welcome to secondary school. If you don't understand anything, just ask. I'll try to make it as interesting as possible.' I was astonished. And even more so when I discovered subjects like Physics and Chemistry. I was over the moon. The science lab became my playground, and I knew that I'd found my vocation. I would have loved to study them at university level, but in those days, money was in short supply and even though my father worked in Guinness's, with four children, there wasn't enough to pay for a third-level institution. However, seeing Daddy trudge off to work every morning to the brewery gave me the impetus to want something different. Whatever I did, I vowed, it would interest and excite me.

So, off I went to Trinity College Dublin to work as a lab technician in the Department of Pharmacology, while studying for A-levels in Physics, Maths and Chemistry in Kevin Street. I have to say, I was as happy as Larry, with my long hair and my sideburns! While in Trinity, I focused on joining every society I could think of, including the Communist Party. I told myself that I was more interested in tearing down the establishment than in being part of it – and there were free pints on a Friday afternoon. But quite honestly, I was having the time of my life being a rebel. It suited me, or so I thought. The army was for other people, like my brother Philip, who was in the FCA, the Irish Defence Forces reserves, and doing brilliantly. It wasn't for Che Guevara here.

Only my mother wasn't impressed. This wasn't the future she'd imagined for me at all. One morning, at home in Rathfarnham, I woke to find her standing over me, a copy of the newspaper in her hand. She started in a roundabout way, talking about our neighbours, Pat O'Brien and Seán Norton, who were army officers and paragons of society, but I had an inkling where she was heading. Finally, she said, 'There's an ad here for the cadets,' brandishing the paper.

'What cadets? The only cadets I know are Eileen Reid and the Cadets,' I joked. The showband was very popular at the time, but Mam didn't smile back. She simply said, 'Look, the ad is in the paper, and I think you should go for it.'

'What would I have to offer them?' I said.

'Well, you're a good organiser,' she said gamely. In fairness, I did enjoy organising events and planning activities, but never had I thought of myself in uniform, not once. Unlike my brother Philip, I had been in the FCA for a very, very short period of time and hadn't exactly covered myself in glory. Eventually, I said, 'Look, Mam, you fill in the application form and send it off.' Anything to keep my mother happy, I told myself.

She did exactly that, filling it in in her immaculate handwriting and sending it off. I forgot completely about it.

Two weeks later, I was invited to attend an interview for cadet training at St Bricin's Military Hospital. I had never thought that I'd actually make it to interview, so I didn't take it that seriously. This had been my mother's bright idea, after all. When she suggested buying me a new suit for the occasion, I demurred, although I did let her buy me a new jacket and consented to wearing a tie.

Off I went during my lunch break in Trinity, having left an experiment to settle, and made my way to the hospital in Arbour Hill. I opened the door of the interview room, to see about eight officers sitting behind a table in full uniform, medals and all. I was invited to sit in the solitary chair in front of them. I swallowed nervously.

'Welcome, Mr Lane,' one of the officers said. 'You might like to take your overcoat off.' And I replied, 'I actually have to get back to Trinity because we're doing an experiment and it's kind of time limited, you know?' When I think of it now …

'Oh,' the colonel said. 'Well, sit down, so.' I answered the questions that I was asked, fairly standard ones about where I was from and what I was studying, and it was all going quite well. A little too well. Still, I reckoned, I had ample time to pull back.

'What have you got to offer the Defence Forces?' I was asked.

I saw my chance. 'Actually,' I replied, 'the reason I'm here today is to find out what you've got to offer me.' There was a long silence, then the colonel looked at the youngest officer at the table, a Lieutenant John Ryan, whom I came to know well and who was a lovely guy. 'Lieutenant Ryan, tell Mr Lane what the Defence Forces have got to offer him.' There was only the slightest trace of sarcasm in the request and John Ryan explained all about camaraderie, *esprit de corps*, overseas service, sports, leadership, dealing with tricky situations and all the rest of it. I listened carefully, thinking that it didn't sound that bad. Not that I'd be joining any time soon.

When John had finished, the colonel looked at me and said, 'Well, Mr Lane?'

I replied, 'I'm a team player, that's for sure. I like to be part of a group and I do like taking charge and leading events … and the

13

thought of overseas service is definitely very interesting ...' Then it occurred to me: was I signing myself up for this? I got hastily to my feet. 'My experiment,' I said. 'I must get back to it.'

There was a sudden commotion as they all stood up, and the colonel said, 'Of course, Mr Lane. Thanks to Trinity for releasing you for the interview. We'll be in touch.'

I ran out the door and jumped on a bus back to Trinity College, but when I got home that night Mam asked me how the interview had gone.

'Mam, I did my best.' I wasn't sure if this was strictly true, but I'd certainly been myself: a bit cheeky, but straightforward, and if they didn't like it, I thought, it was their loss.

A few weeks later, the brown envelope arrived. It was marked 'An Roinn Cosanta' and the letter inside was in Irish. I was in bed after a night out, and Mam brought it up to me in a state of great excitement. 'You read it,' I said blearily. She looked at me and said, 'What are we going to do with you?' And she picked up the envelope and stomped back downstairs to listen to the *Gay Byrne Show* on RTÉ Radio 1.

Five minutes later, I heard her running back up the stairs. She burst in through my bedroom door. 'You've to report to the Curragh in June.' I couldn't say anything. I was in a state of shock.

CHAPTER 2

MAKE OR BREAK

began my two-year officer training course at the Curragh on the understanding that I would stay for two weeks. If, at that point, it wasn't for me, that would be it, I'd told my father. Like a sensible man, he agreed. Parents are psychological experts and despite being a bit sceptical about the Irish Defence Forces, my father was as keen as Mammy for me to do something with my life.

My parents accompanied me to the Curragh on that first day in June 1973 to join the 49th cadet class. Who came over to greet us only Lieutenant Ryan, the man who had told me about what a great place the Irish Army was. I was dressed like any other man in the 1970s, with hair down to my shoulders. His was neatly cropped and he was in full dress uniform, shiny Sam Browne belt across his chest, clutching his hat. My mother could hardly contain herself. 'Oh my God, are you going to be like that? Imagine,' she said, awed by the spectacle. I was thinking, *I'm out of here in two weeks*.

My first rude awakening came when it was time to get a haircut. As I had long hair and sideburns, when the barber said to me, 'What way would you like your hair?' I instructed him to keep the sideburns and just take an inch or two off the end. Out came the shaver and he proceeded to shave the whole lot off. I now had a short back and sides. I couldn't believe it.

A week or so later, I'd been released from the Curragh to do one of my science exams, and I can still remember standing under the clock at Clery's department store, waiting for a girl I was seeing at the time. She walked right by me, failing to recognise me in my new haircut. I can remember standing at the clock, thinking that the only place that I could feel safe with my new haircut was in the Curragh. I didn't fit into 'regular society' anymore. So, I got back on the half-ten bus, which arrived at the Curragh at five to twelve, and that was the start of 45 years.

◆

Unlike at primary school, I knew that this was a game I wanted to play. I learned very quickly. The mantra seemed to be 'minimum sleep, minimum food, maximum exercise'. We'd get to sleep at one in the morning, and at 3 a.m. we'd be woken by a voice on the tannoy: 'All cadets on parade outside.' We'd have to get up, get our combat uniforms on, and out we'd go, all 52 of us, to stand there in the darkness. I couldn't see John Ryan, who was now my platoon officer, but I knew that he was there in the dark, watching. Now and then, I'd see his hat or badge catch the light and then he'd call over, 'Cadets, dismissed.' And we'd go back to bed. Twenty minutes later, we'd be back down again. And this would go on and on.

One famous Friday afternoon, I saw four of my cadet class go in to see Captain McKevitt, the CO (commanding officer), and they resigned. They'd had enough of the endless activity and the lack of food and rest. We were due a weekend pass that weekend and we all thought, well, they'll have to go easy on us now. They're bound to feel sorry for the rest of us now that four of us have dropped out! Five o'clock came and the announcement came on the tannoy: 'All cadets into Seomra An Phiarsaigh [the briefing room]. Here we go, we all thought, in a state of high excitement at the prospect of 48 hours' freedom. But when we arrived, we were told, 'Into combats. Obstacle course.' The weekend was gone.

After a while, I realised that there was no badness in this method, and all they were trying to do was to break us: to put it simply, to put us under pressure to see how we'd perform. I found that I had no problem in following the rules that came with my cadet training. Just like I had in primary school, I said to myself, 'They're not going to get on top of me. They're not going to beat me.' But this time it was different. I knew it wasn't personal, even if I was on the very edge of exhaustion all the time. And I loved the challenge.

We also had classes as part of our training, and if we were caught nodding off, the officer would come up and tap one of us on the shoulder and say, 'Cadet, what have we just done?'

'I don't know, sir.'

'Why did you fall asleep?'

'I don't know, sir.'

'What do you mean, you don't know? Is it because you're not getting enough sleep?'

'Oh, no, sir.'

'Okay, guard room, three, half-past three, four o'clock.' We all knew what that meant. We'd have to wake at those times, get up from our bunks, go to the guard room and sign our names and the time into a ledger, then go back to bed and try to go back to sleep. I knew that this was harmless enough – because if they *really* wanted to get you, they'd get you to run to the Curragh racecourse and back every half an hour for a couple of hours! I also surprised myself by enjoying some of the classes, particularly one on leadership, tactics and military history. I found that my busy mind was fully occupied, and exhaustion dealt with the rest.

I also learned my first real lesson in life: that leadership wasn't about shouting orders or, indeed, accepting orders blindly and without question. It was about helping your unit to function as a group with everyone's opinions being counted. Our cadet class was full of characters, but we all pulled together to get everyone over the line. I took this lesson with me into my career in bomb disposal: I always checked with my team that everyone was happy with the chosen direction, although I was clear that, ultimately, the responsibility would be mine. That was what cadet training was for: we would be the leaders of the future, so we needed to prove ourselves.

It wasn't all grind: I was part of the cadet school debating team participating in the *Irish Times* debating competition, along with some very gifted talkers. I was surprised to find myself outclassed in that department. I qualified as a club diver in scuba diving in the force's sub-aqua club, which ensured I had some well-earned time off in the second year for my final assessment.

Our first year in cadet school seemed to go on for ever but finally, the following June, we went to the Glen of Imaal in remote Co. Wicklow to do our final exercises. Looking back at the class yearbook recently, I came across an essay on the exercises by Mick Delaney. 'We looked upon it in awe, as a God, something to be feared and at the same time, respected.' We did. We'd heard all the stories about this fabled set of exercises. As officer cadets, we'd be put in charge of a platoon of maybe thirty or so of our own people. Scenarios would then be set for us: offensive operations, defensive operations and counter-insurgency. In the offensive exercises, we would lead our platoons in an attack; in the defensive, we'd dig in for the duration, which is no joke; and in counter-insurgency, we'd be dropped off in the middle of nowhere and told that we were being followed for the next three days. I loved the adrenaline and the excitement of that exercise. Captain Des Travers had put together a series of night patrols in the woods, which really tested us.

We were simply humble cadets, there to be pushed to the limit. So, the very worst of our exercises was saved for last, an exercise known as 'scratch'. It might seem simple, but at this stage, after three weeks of exercises, we were totally exhausted. We'd be assembled in a field, and we'd have to run around it, then we'd be stopped and asked for our name and number *as Gaeilge*, and if we missed it, they'd send us around again. It's called 'scratch' because you'll end up scratching the ground with exhaustion. Finally, we would walk the 30 kilometres from the Glen of Imaal to the Curragh with full packs.

I can still remember my first-year tests, because on the way back, all I could hear was my feet squelching in my boots. I thought it was mud, but then I realised that my heels had been cut to ribbons and my boots were full of blood. But I also knew that stopping at this stage would be catastrophic, so I kept on going, limping all the way back to the Curragh. Eventually, we got back, and we all assembled in the exercise yard, standing as upright as we possibly could. Lieutenant John Ryan came out. 'A poor effort by all' was his usual comment, followed by 'Is there anybody here who has to go to the hospital?' Nobody said a word. My boots would have to be cut off me, but I'd rather they were than to speak up. 'Okay, fall out, you're off for the weekend.'

I sneaked off to the doctor to see about my heels. He took one look at them and said, 'I'm reporting this. Who's in charge of this cadet class?' He was absolutely outraged, ranting on about us being treated worse than animals. I thought, *please just fix it*. I knew that if I made a complaint, it would only make things worse in the long run. He bandaged it all up and that was it. I'd managed to survive the first year of my cadet training.

Of the 52 of us who started the class, minus the four who had resigned, half were to pass out (qualify) at this point, and the other half were going to university, so we were to stay on for a second year. In the Irish Defence Forces, specialists in bomb disposal, engineering, logistics and physical education have specific qualifications obtained from many different institutions, in my case UCG. I couldn't wait to get started. One of the great benefits of being in the Defence Forces was that our higher education was paid for, on the understanding that we'd be using it to do our jobs later on. They were spending money on us now to reap the rewards in the future.

The great day came when half of us were being commissioned. Out came John Ryan onto the square and he fixed us with a glare and said, 'Where are the half who are going to university? Right, I want you to fall out over there.' Then he continued, 'Into your combats and onto the obstacle course.' The rest of the group was to assemble in full dress uniform for the passing-out ceremony, while we were dragging ourselves once more over the obstacle course. The idea was, if you lot think you're ready to be real soldiers, there's no chance. I got the message. WILCO.

CHAPTER 3

THE BIG BANG

was commissioned in 1975, in the teeth of the Troubles, and the conflict defined my early career in the Defence Forces. The Dublin and Monaghan bombings had happened just one year before, in May 1974, and as a young officer now training in bomb disposal, I had been marked by them. In fact, I'd been in Heuston Station with the army sub-aqua team, setting off for a trip to West Cork, when the news came through. Three car bombs had gone off in rush-hour city-centre Dublin, followed by a bomb in Monaghan just 90 minutes later, killing 34 people. The bomb killed one entire family, the O'Briens, a mother, father and two very young children, who just happened to be walking by the car when the bomb exploded. The photos of flames shooting up into the air, windows blown out and wreckage strewn all over the streets that appeared in the newspapers in the following days were hard to shake. I later learned that ordnance officers had gone out to render safe a number of suspect devices, many of them with no suitable equipment – incredible bravery that made me want to join the corps.

I was particularly struck by the fact that much of the forensic evidence was lost in the haste to sweep up the damage and to get the streets open again. Nowadays, collecting forensic evidence is a key part of our job, and I can't help wondering if the perpetrators would have been found had this been the case in 1974. As it is, even though the Ulster Volunteer Force (UVF) would ultimately claim responsibility, no one has ever been charged with the bombings, in spite of all the investigations and reports. What I found curious was that the day before the bombings, the UVF had no capability to undertake this level of complex attack, and the day after, they also had no capability. I couldn't help wondering if they had been helped along the way. Indeed, according to the CAIN Archive, Ulster University's library of information on the Troubles, 'While no firm conclusions were reached, it was suggested that the security forces in Northern Ireland were the most likely source of help.'

By this stage, I was in my second year of cadet training and loving student life in UCG, thinking that I cut a bit of a dash in my uniform! I also had a car, which was a bit special back then. I had initially enrolled in Archaeology, which baffled my CO. 'I can't see anything in your Leaving Cert about archaeology.' I didn't want to tell him the real reason: someone had told me that there were lots of women in Archaeology. Thankfully, before too long, I saw the light. I met my wife, Bridget, at a party in Dublin – she rightly ignored me at first – and I soon transferred to the science block, where I could develop my passion for chemistry, knowing that it would be put to direct use in my job. By this stage, I was eyeing up the possibility of a role in the Defence Force's Ordnance Corps, where I knew that science expertise would be essential.

My mother finally got her heart's desire when I was commissioned with the rank of second lieutenant, the entry-level rank of a commissioned officer in the Irish Defence Forces. (The Irish Defence Forces have two streams of soldier. Non-commissioned recruits start as a private and work their way up to corporal, then sergeant, all the way up to battalion sergeant major, offering support to the commissioned officers, who have had specialist training. *They* begin as second lieutenants, then progress to lieutenant, captain, commandant, colonel, all the way up to lieutenant general, the army's chief of staff.) Just like John Ryan four years earlier, I stood in my officer's uniform with my Wilkinson sword and Sam Browne belt and pledged my allegiance to the Irish Constitution. I received my first pip on my shoulder and my first posting to the 12th Infantry Battalion in Limerick.

I was to be based in Sarsfield Barracks with my good friend Paul Rossiter, where I met some unbelievable characters, including a man called Des Rooney, who featured in the works of Brendan Behan. 'The fat captain at Harold's Cross' was the writer's description of Rooney. By the time I arrived in Limerick, Des was a seasoned pro and too long in the tooth to be dealing with another bunch of eager officers, still wet behind the ears. However, he also had a passion for greyhound racing, which elevated me in his regard. One day, he was sitting beside us at lunch reading the *Limerick Leader*. His head shot up and he fixed me with a look. 'Is this you?' he said, brandishing the paper. There I was in a photo in the sports pages, receiving a trophy at Limerick greyhound track! I had no intention of putting him straight and telling him that I had no interest in greyhounds: the dog in question was my brother-in-law's and we were only there to support him. From then on, Des had me plagued!

In Limerick, the first thing I had to navigate was my new-found status as the leader of my own platoon of 30 soldiers, many of whom were older or more experienced than I was. I was the new house doctor, full of it, but a bit green. I had the qualifications on paper, but some NCOs – non-commissioned officers – would look at me and I knew what they were thinking: *he's barely out of school, what the hell does he know?* My job was to show them what I was made of, bringing them to the Border for exercises and to Portlaoise Prison to provide security. As an officer, you can empathise with your troops, but you can't be one of the lads, because in the heat of battle, you are the one who will be giving the orders. Your job is to grow a team around you, to show them that they can rely on you to make the right calls when needed.

I had great fun with my platoon, but I quickly knew that the infantry wasn't for me, so two years later, in 1978, when the call came for the Ordnance Corps, I was the first to put my hand up. I wanted to be a team player, but I also wanted a job that would tax my brain in a different way and that would allow me to use my science degree. For those of you who don't know, the Irish Defence Forces is split into a number of 'corps': Infantry, Artillery, Cavalry, Engineer, Communications and Information Services, Ordnance, Medical and Transport. Put simply, infantry are the soldiers on the ground; artillery use equipment to fire on the enemy; cavalry drive tanks and other equipment; engineers ensure safe transport of the army, clearing obstacles to progress, etc.; communications provides IT support and services; medical involves providing medical and dental care to members of the Defence Forces; and transport cater for moving army personnel safely between locations and providing logistical support.

The Ordnance Corps has two missions in the Defence Forces: logistical and operational. According to our remit as described on military.ie, 'The logistical role of the Ordnance Corps is to provide technical support to the Defence Forces for the Procurement,

Storage, Distribution, Inspection, Maintenance, Repair and Disposal of all items of Ordnance-related equipment. The operational role of the Ordnance Corps is to train personnel for and provide the State's EOD [explosive ordnance disposal]/IEDD [improvised explosive device disposal] capability.' So, it was our job to learn everything about every weapon in the Irish Defence Forces' arsenal, including in the Navy and Air Corps, as well as preparing for our bomb-disposal validation course. The corps is also responsible for everything to do with catering and uniforms, a role which I happily took up a few years later in Lebanon.

The ordnance training course lasted just over a year and my CO was Commandant Murt Clancy. You quickly learn that in the Ordnance School, you can undertake a huge range of courses on everything from armourers' training to homemade explosives, the components of a bomb, from the time and power unit, or TPU, containing the timer and batteries which supply the explosive charge, to the various kinds of explosives we might encounter, from fertiliser to icing sugar, and the chemical reactions that would make a bomb actually work. We learned that every bomb contains some means of initiation, an explosive or incendiary charge which will either explode or catch fire, one for detonation, the other for burning; an arming/safe mechanism, for example a switch – because as the bomb is transported, you don't want it going off. When the location for the bomb is reached, the safety switch is turned off and the arming mechanism is turned on. The bomb is now activated.

The Provisional IRA were experts in the car bomb, but the British Army became proficient in intercepting them. So, in later car bombs planted by the IRA, an incendiary mix was placed

29

into the container with a timer; when the timer was initiated, the mix started to burn, giving off white smoke, which surrounded the vehicle, giving the British Army no time to intercept it. Ten minutes later, the vehicle would explode. Training is the key in bomb disposal, not flair.

My training would eventually take me back to the Glen of Imaal, where I was presented with a scenario to test my abilities. I'd be marked on the outcome, but also on each step of the process: in logical thinking, understanding the process and not taking risks unless I could justify it, among other criteria. I can remember the assessment as if it were yesterday. In one scenario, we were given the task of dismantling a – hypothetical – bomb in the control room of a hydroelectric plant. It was hard to imagine, standing in a wet field in the Glen of Imaal, but we had to try.

Our DS (Directive Staff), Lieutenant Brian Flood, had asked us to imagine that the terrorists had gone up to the top of the thirty or so steps of the plant, opened the swing door at the top, and placed the bomb in the control room before running back down. The lights wouldn't work, so we concluded that we were dealing with some kind of light-initiated device. Next, one of us was to play the role of a hungry sniffer dog, complete with light around his neck! Meat was thrown onto the stairs and the 'dog' ran up to get it, the idea being that it would encourage him to head to the control room. However, in the scenario, the guy who was playing the dog decided to throw a curveball: he got two thirds of the way up, had plenty of meat and came back down again, full. 'What's your next step, Lieutenant?' was the question. Then we were told that, by the way, the bomb was to go off in 10 minutes.

Of course, this was just a scenario, but we all knew that if a bomb was to go off in 10 minutes, we had to do something. In my colleague's case, he got into his bomb suit, put his weapons system together and disrupted the bomb's time and power unit as he'd been trained. We lived to fight another day.

I can remember that at the time, I found it all a bit unlikely, because, of course, there *was* no hydroelectric plant, or car, or aeroplane. When I went on to train in the UK, I was very impressed with the entire town that they provided for our exercises, including an 'airport', shopping centre and so on. In fact, when I joined the Ordnance School as an instructor some years later, I made sure that the exercises were in real-life situations, such as shopping centres, Dublin Port, Dublin Airport, petrol trucks, trains – anywhere a bomb could be placed – so that the officers could properly test themselves.

One example comes to mind of the importance of real-world exercises. A good friend of mine, a laid-back guy, was being assessed for his competence and was given the scenario of a petrol tanker with a bomb on it. The tanker was full of fuel, so the question was what was he going to do – would he get into his bomb suit, or would it hinder his operation? So, he had to justify that. He couldn't climb up the steps of the petrol tanker with a bomb suit on, and the bomb was out of reach of the robot, but if he wasn't going to wear his suit, could he justify it?

He decided to get a full sense of how petrol tankers worked, how the fuel could be unloaded and what kind of tools were needed; if the petrol was removed first, and the bomb went off, only the tanker would be damaged – could he live with that? However, if it went off with 50,000 litres of fuel on it, that would be a different scenario altogether.

He got the driver over to quiz him and was told that there was a quick-release tool to empty the fuel, but it would take 90 minutes to drain the tanker. The timer on the bomb was set for 35 minutes. At that time the device would function, causing the petrol tanker to detonate, leading to significant damage and potential loss of life. My colleague was going through the whole process on a whiteboard in the back of the bomb-disposal vehicle, using the three criteria we always apply: worst case, best case, most likely. Finally, he said, 'I'm going to partially wear the suit, helmet, flak jacket, nothing else. I want a disruptor tool loaded and I'm going to render safe the device.' (A disruptor is a tool that fires water at supersonic velocity to disrupt the firing mechanism of the bomb.) He had managed to get a robot a certain part of the way up the tanker, and he could just about see the edge of the bag containing the bomb. He'd X-rayed it and got a sense of its size and whether the disruptor would work. His whole plan was coming together. He looked at his number two and asked him if there was anything he hadn't thought of. The answer was no, so he was good to go.

In my test, I was presented with a car in which an improvised mortar bomb had been placed in the boot. The mortar bomb came complete with a TPU. We were under serious pressure, so the first thing I decided to do was to attack the TPU, which I did successfully using the disruptor, and then I went at the improvised mortar, using the robot to lift it out of the boot onto the ground. So now the TPU had been neutralised, there was no power to the mortar, and it was safe. Job done.

I took the evidence and bagged it for forensics, then I was ready to hand over the scene to the Gardaí. The 'Garda' walked up to the back of the car, looked in and put his hand under the seat, taking out the detonator. When I had fired my weapon into the TPU, the detonator had blown under the seat, and I hadn't searched the car thoroughly enough afterwards. A clear fail and rightly so. However, the worst word you can hear is 'fail', because the whole process takes hours. In the case of the car bomb, it had been *in situ* for hours; there was no threat around it, the area was evacuated and I'd gone through all my procedures and thought I'd done brilliantly – but I should have identified that I hadn't retrieved the detonator, because I'd X-rayed the bomb earlier and seen the detonator!

The Ordnance Corps training is very exacting, because it has to be. Every step that is laid out in our *Tactical and Technical Aide Memoire* has to be observed and we are marked on that. The *aide memoire* covers every possible technique and piece of equipment that we might come across. If we fail to observe a step or do something different, the question is why? If we can provide a reason why we've come to a particular decision or taken a particular action, that's fine. Otherwise, it's goodbye. We get one opportunity to fail and that's it. As we are all qualified scientists or engineers, high standards are expected of us, but we are also allowed greater flexibility than in, say, the British system, where bomb-disposal experts are well trained but not necessarily specialists. Our system is expertise-driven, rather than process-driven, and that's a powerful combination.

I loved the mental challenge of the bomb-disposal officers' course and the technical demands of the ammunition course, in which you strip weapons down to the nuts and bolts, learn how to replace parts and, crucially, how to prevent weapons-related accidents. I found that I could use my busy brain to dismantle complex systems, to see how weapons and bombs actually functioned and what might have been going on in the bomber's mind when a device had been made. For me, it was the perfect outlet for my curiosity and my desire to work things out – I sometimes wonder what things might be like if we had more of this kind of learning in school. We also covered the nascent area of CBRN (chemical, biological, radiological and nuclear weapons) and how they could be integrated into improvised devices by terrorists. It was fantastic, but tough going. There was so much to learn, as well as the pressure of having to pass my assessments.

At this time, I was instructed by and met many fellow ordnance officers with real professionalism and moral courage. These were men who in 1972–4, with limited equipment, went onto the streets of Dublin and other places to render explosive devices safe while protecting the public. These men included D.K. Boyle, Harry McGennis, Joe O'Sullivan, Denis Ward, Des Donagh, Murt Clancy, Jimmy Barret, Rory Kelleher and many more whom I did not work directly with. It is a source of surprise to me that no official recognition was ever given to the bravery shown by these people during very difficult times.

Then I was off to Kineton, near Stratford-upon-Avon, to train with the British Army on their specialist course in bomb disposal. It was called the ATO, or Ammunition Technical Officer's course. As part of the course, we were to take every type of ammunition in the NATO inventory and learn how to break it down, strip it down, its shelf life, how to do chemical testing on it. We were to spend lots of time in laboratories doing experiments, so I was in my element. There was also the promise of small- and large-scale demolitions, blowing up cars and beer kegs and things like that, which is every boy's dream! It looked like just the job for me. I was full of myself and my hopes for my career: little did I know that it might be derailed when it had barely started.

CHAPTER 4

KINETON CALLING

We arrived in Kineton on a bitterly cold day in January 1983. It was a difficult time to be Irish in Britain, but I was initially surprised at the lack of anti-Irish sentiment in the British Army. They were losing guys in Northern Ireland, but with the exception of one or two of them, the welcome to me and Bridget and our newborn, David, was warm and helpful. Of course, there was the odd comment, a question about whether I was in the IRA, to which I replied, 'If I was, do you think I'd be here?!'

I can still remember the deathly cold in the house that we were allocated when we opened the front door. It later transpired that the information I'd been sent about the course – and hadn't read, because, apart from in my job, I'm not a great planner – had warned me to order coal in advance. Bridget was not impressed. 'Ray, I'm telling you now what I'm doing tomorrow. I'm getting back in the car and going home. This is ridiculous, with a baby.'

She was right. Then the doorbell rang. *Saved by the bell*, I thought, going to open it.

A man was standing there in uniform, hat on his head. 'You're from the Republic of Ireland? Captain Lane, correct?'

We shook hands. 'Ron,' he introduced himself. He was an officer associated with the course and we were to become firm friends. 'Do you have any heating or food?'

'No,' I replied. Bridget appeared beside me and I introduced her. 'This is my wife; she's going home tomorrow.'

'I'm not surprised,' he said. 'Hold on, we'll sort it.' A few minutes later, he returned with a bag of coal and his wife came in with food, and the house started to get warm and things began to look a bit brighter. Bridget was looking a bit happier at this point.

We were sitting around chatting when Ron took his hat off. It was all I could do not to gasp, because his face had been completely melted away – it looked as if it were made out of plastic. It turned out that he'd been blown up in a car bomb in Northern Ireland. To this day, I often think, if the roles were reversed, would I have been as nice to him as he was to me? I doubt it. We in the Defence Forces co-operated with our British counterparts, but when it came to Northern Ireland, that was very much their domain.

Thanks to Ron, we began to settle into life in the British Army, which was quite a bit more structured than in the Defence Forces. Bridget was surprised but happy to see that she was expected to play a full part in army life and quickly settled into the many activities provided for army spouses. The regimental dinners were a particular high point, with all of the ritual and dress uniforms and so on, but at the end of which mayhem was always guaranteed. I became notorious for being the quickest to do a 60-yard dash over the furniture in the mess hall, managing to get myself into the bad books of the colonel who ran the school in the process. I can still remember returning home at six o'clock in the morning after my triumph and saying to Bridget, 'Whatever happens, wake me at seven,' because school started at eight. And I slept it out. Like a bold child late for class, I tried to sneak in the back door of the school, but it was locked so I walked around to the front and who should be waiting for me? The colonel, and he'd been there waiting for me for 40 minutes.

I walked by and gave him the Irish salute, which is slightly different to the British one, which probably added insult to injury. (In the Irish Defence Forces, we are taught to salute 'longest way up, shortest way down'. We extend our right arm out, then bend it so that our fingers rest, facing down, at the edge of our hat. In the British Army, the bent arm is lifted upwards to the hat, palm facing out.)

He gave me the British salute and I said, 'Good morning, sir,' and got no reply. Off I went into the classroom, head hanging off me. The next thing, the door opened and Major Alan Morley, the officer in charge on the course, stuck his head around it. 'Ray, the boss wants to see you.'

As we walked towards the colonel's office, he warned me, 'We'll go in here and you keep your mouth closed. Just shut up and don't say anything. Let him get it off his chest.' I've never been one to remain silent, but this time, I vowed to take Alan's advice and do what I was told.

When I went into his office, he said, 'I'm very unhappy with what I saw last night in the officers' mess. I'm also very unhappy with your general performance here.' I couldn't help myself. I was about to say something in reply, but Alan was behind me and kicked me with his foot. And then he added, 'And I know your wife really hasn't settled in either.'

I knew that Alan was right and that I needed to keep my mouth shut, but at the mention of Bridget, who had been so happy and who had thrown herself into everything, I saw red. 'Sir, you can say anything you like to me, you can make me climb that flagpole out there and kiss the Union Jack, because you are my commanding officer, but you will not talk about my wife.'

At this, the colonel lost it. 'Get out and go home to your little country. Out!'

I saluted him and walked out the door and I went home to Bridget. I said, 'I have good news and bad news.'

Bridget rolled her eyes to Heaven. 'Tell me the good news, then.'

'Well, there is no good news. We're being sent home.'

'What? What did you do?' I had to tell her about the 60-yard dash over the mess-hall furniture and my subsequent encounter with the colonel. It took another visit and the smoothing of ruffled feathers by Alan to restore our understanding. I knew that the after-dinner games were very much part of British Army life, so the unfairness of being singled out smarted a bit. However, when I learned that the colonel had lost people in Northern Ireland, I resolved not to nurse any grudges. I got it. In his position, I might well have done the same.

I'd really been looking forward to training with the British Army because their level of expertise in bomb disposal at that time was the global standard. My ambition at that stage was to make Ireland the global standard, so that was why I was on the course: to see what we could learn and to bring the lessons home to Ireland. With so much activity on our side of the Border with Northern Ireland, it seemed more important than ever that we learn from the very best. The man in charge of the training was a Staff Sergeant DS. He would be assessing me and marking me as either 'pass' or 'fail', based on my performance in a number of hypothetical scenarios. Unlike in Ireland, the British Army had a purpose-built village, the Felix Centre, at Kineton, with an airport, fire station, a supermarket, everywhere somebody might think to put a bomb.

I had been flying through the course, passing every test and acing the technical exams. Now, for our first task, we were on the Border – hypothetically – looking into the Republic of Ireland, i.e. at friendly forces. When I mentioned this, the DS fixed me with a look and said, 'Well, that's to be debated.' So, I knew I had a problem straight away. I knew that not every member of the British Army was going to look on me with warmth, but I had been more than pleasantly surprised at the welcome I'd received so far. How on earth, I thought, had I managed to stumble across one of the very few exceptions to the rule? I also knew that I would have to get on with it.

The very first scenario was six hours in length, involving defusing a bomb on a train. I knew the drill: you go through a full risk assessment, you consider your options and you look at the potential results of any action you take under our three headings: the best, worst and most likely outcomes. I was to design my procedures around that and to ask questions of the DS accordingly. If you don't ask, you won't know what you need to know, so in this case, my questions might go like this:

'What's on the train?'

The answer might be 'A thousand litres of petrol.'

'Where's the petrol?'

'In the tank behind the driver's console.'

'Has the train been evacuated?'

'Yes.'

'What area around the train have you evacuated?'

'A hundred metres.'

I would then consult my *aide memoire*, which I'd annotated heavily during my training, and say, 'No, I need more space around the train.' I double-checked the manual, then continued to work out the parameters: 'Are there any trains due on this line?'

'Yes, there's one coming in 45 minutes.'

'Stop it,' I ordered. 'Are there any planes flying over?'

'Yes, there's an airport over there, as you can see.'

'Okay. No flights in or out.'

I went through every single option, then I went into action. I had the train stopped, I had the place evacuated, I had emergency services on call, I had ambulances, fire tenders, I had it all there. Then I proceeded to get as much information as I could about the bomb. I eventually identified where in the engine the bomb was and who of the different actors – the train driver, the head of security and so on – would answer any questions I had about the bomb accurately. I was putting the whole thing together in my own mind. I asked the driver if there was access for my robot, to which he replied, 'What's a robot?' I showed him, and then asked if I could get it on the train. Not in a million years, I was told. So, I got the construction blueprints of the train and confirmed that I couldn't put the robot in, so I'd have to take a manual approach. Not ideal.

When we start our training, we learn that we can take one of three approaches to a scene. The ideal is remote, of course, where the robot can be sent in while we monitor on our screens, then semi-remote – i.e. if I can't get the robot in, can I get it to a distance that may be of value to me? The last resort is manual, because this poses the greatest risk. And, as bomb-disposal experts, we are all about reducing risk. Having said that, there was a famous – and possibly apocryphal – case of a bomb being left in a Belfast hospital ward during the Troubles. When called to defuse it, the bomb-disposal team simply grabbed it and threw it out the window. Why, when they could have blown themselves and half the ward up? Well, there was clearly no way of getting a robot in, or getting it to any meaningful position, and crucially, there was an immediate threat to life. An immediate threat to life means, quite simply, that the gloves are off and you take the quickest approach to getting the job done.

Back to my exercise. I went around the train with the robot, to get a 360-degree view of where I was operating. I was able to ask the train driver, 'Tell me how you got off the train,' and he said, 'I got down those steps there.' So, I knew that it was safe to go up the same steps. I donned my suit, and I went in. The diagram of the device that the driver had given me was perfect, so I had put my own plan together on how to attack the device using a disruptor weapon system. We are taught to take the longest way through the device through its weakest point: the longest way through means that whatever's in there, you're going to get it and hopefully hold it for forensic evidence rather than blowing your way through it. And the weakest point for obvious reasons.

So, in I went, happy with my decision, then I returned to the vehicle in my suit, sweat pouring out of me. I gave the mandatory warning: 'Firing now. Everybody okay, everybody under cover?' When I got the positive response, I fired the disruptor. It made a loud bang, then, after the mandatory soak time, I went back to see what was left of the device. The detonator had come away from the battery and the battery had been disrupted, so I put everything into forensic bags and brought it out to hand it over to the police. Job over.

The concentration required was immense, as you can imagine, so I was exhausted, but happy with the way things had gone in the first of my tests.

The staff sergeant appeared and said, 'Captain Lane, how do you feel this job went?'

'I'm very happy with it. I thought we came up with a very good RSP.' (We call the process a 'render-safe procedure'.) 'Obviously, safety was paramount and nobody was affected, there was limited damage to the train and the tracks are now open. Yeah, I'm extremely happy with it.'

Again, I got the frosty stare. '*You* might be,' he said. And I just knew that I was going to fail the task. The sergeant waffled on, but I wasn't listening to him, because I knew that I'd done a good job, and because I was concentrating on the final score to see whether I'd passed or failed. 'And that's a fail,' he said. 'Have you any comment?'

'No,' I said. I was devastated, because I knew that of the nine tasks I was allocated, I could only fail one, and this was it. I couldn't afford to fail another.

Throughout the week, observed by different instructors, I dealt with various scenarios of bombs in shopping malls, football grounds, etc., and passed all of them with flying colours. Then I came to the ninth and final task, which was to defuse a bomb on an aeroplane. I knew that I simply had to pass it. I was a bundle of nerves as I prepared for the day, hoping against hope that I'd pass.

The scenario that I was presented with was as follows: there was a bomb in the cockpit of the aeroplane. In the passengers' rush to get out, two of them had broken their legs and were trapped inside. The pilot had been shot, although whether they were still alive, we didn't know. The device was under the co-pilot's seat, but he had got out. The DS for this task was the very same one who had examined me and failed me on my first task. I knew I was in trouble even before I started.

I began my process with a question. 'Do you have any more information on the explosive device?' The answer was that it was a timed device and it was going to go off in 55 minutes, starting from that moment. I began working through possibilities on the whiteboard, before saying, 'Okay, we'll get the people with the broken legs out. That'll only take ten minutes.' They had two people who had volunteered to have broken legs and lifting them out was some task, as they were not helping!

I couldn't help the pilot, because they were unconscious, slumped over the controls. So, I went into the cockpit, and I X-rayed the bomb using our mobile X-ray and got a brilliant image. I could see the whole device in front of me: the detonator, the electronic timer, the whole thing. I knew what I needed to do next. I got hold of the disruptor and prepared to fire into the timer without causing any other damage. This was essential, because remember, the captain was lying injured only feet away. The needle shot into the timer and destroyed it, rendering the bomb useless. I breathed a sigh of relief.

The sergeant appeared beside me. 'Well, Captain Lane, what do you think yourself?'

I had an ominous feeling that it didn't much matter. I blurted out, 'Staff Sergeant, let me tell you what *you* think.'

'You can put "Sir" before that if you're addressing me again.'

I ploughed on. 'That operation went extremely well. The pilot was saved, everybody was saved, there was no damage to the aircraft. Now, I know you're going to have a different viewpoint on that.'

'Exactly,' he said. He jabbed a finger at the bottom of the page, where I could see the word 'fail', written in block letters. I stared at them, knowing that I had a choice. I could fail the course and go home, tail between my legs, or I could call the sergeant out on whatever grudge he was nursing. I had a feeling that my Irishness wasn't helping me, but clearly, that wasn't something I could say out loud. Nonetheless, I knew that I had to say my piece. I'll tell you why.

Before I had set foot in Kineton, I'd happened to come across a BBC documentary, *Nationwide*, following one young British officer who was training to be a bomb-disposal expert. They filmed everything, including his tests at the Felix Centre. And the poor guy failed them. His assessor was none other than the same staff sergeant. Now, testing bomb disposal isn't a subjective process – sure, there are different approaches to the same problem, but if you are following the correct procedures and the outcome is positive, that's not a fail in my book. I remember being really annoyed when I saw the TV programme, having no idea that the very same man would be testing me.

It was do or die, I reckoned now. This wasn't a spat with my colonel in which, arguably, I could have kept my mouth shut.

My career was at stake here and I knew that I'd done the exercise properly. So, in front of the assembled cast, I insisted that they call my main contact on the course, Alan. While we waited for him to appear, we all stood there in complete silence. Staff Sergeant DS was looking daggers at me, but I had to ignore him, because I knew I couldn't give in. Finally, Alan arrived and I said, 'Alan, it's as simple as this. Day one, task one, he failed me. The evenings of days two, three, four and five, I averaged 95 per cent and got outstanding on all of the next seven tasks, then we come to task nine and I failed it. Now, the question is, is it me? Or is it him? And that's your job, Alan, to make that decision.' The reason I got so hot under the collar was that I felt that I'd done a good job. If I'd messed up, I'd say, 'Okay, I messed up,' but I didn't.

The guy was gone out of the school in two days. I wasn't happy about it in the way you might expect. It wasn't about getting the better of him, it was about dealing with the bigotry and prejudice that he represented. Perhaps there was an element of anti-Irishness to it, but I can still remember the way in which he'd ruined that officer's career on the TV programme. To this day, I can still recall the last scene with the poor guy sweating in his bomb suit and the sergeant saying to him, 'Well, the good news is, Captain, that you won't have to worry about this again because you've failed.'

Maybe the Irish flag on my badge had helped me in some way, because I wasn't cowed by the sergeant. I knew that I'd done a good job and I felt the unfairness keenly. The words of the colonel at cadet school returned to me: 'Never do anything without knowing the reason why.' That's all that I'd done.

CHAPTER 5

TRAINING DAYS

Nobody likes to be a rookie, no matter what the field, but we all have to start somewhere, and so I began life as a real-life bomb-disposal officer in Dublin in 1979 with the Papal visit. After all my training, I was dying to get out there and apply everything I'd learned over the previous few years to real-life situations. However, I wasn't let loose immediately, even with my extensive training and practice. First, I had to act as an understudy to an experienced bomb-disposal officer in Dublin, just to see what I shouldn't do! I was to accompany Captain Paddy Boyle, a lovely man.

We were in Clancy Barracks when the callout came. We got the initial information that there was a bomb in the toilets of the Clarence Hotel. At this point, the 'tasking authority', as our handbook puts it, tells us what category level the incident is. We class them from A – 'a grave and immediate threat', to B, 'an EOD incident which constitutes an indirect threat', down to C – which represents a minor threat – and D, where no threat is presented at the time. Most incidents fall into Category B – even at the height of the Troubles, the threat wasn't 'immediate', even if it was serious.

I was both excited and nervous at the same time, but the thing about bomb disposal is that you click into gear the minute that phone rings. The steps that we need to follow, and the procedure, ensure that we keep our heads. As we drove through the busy city streets, I repeated the philosophy I'd learned at the Ordnance School in my head. 'One, preservation of life; two, removal of the threat; three, limit damage to property; four, preserve evidence; five, get things back to normal as soon as safely possible.' As long as I kept them front and centre of my mind, I'd be okay.

Captain Boyle, myself and the driver of our vehicle arrived at the rear entrance to the hotel, around the corner from the quays, which were by now choked with traffic. We evacuated the entire hotel, even though some of the guests weren't happy about it. We closed the road down, then we got a detailed briefing from the Gardaí. Paddy asked the questions: who, what, where, why, when? Who saw what? How long was it there? Why would it be there? All the questions that we'd need to ask to get a sense of the situation. He wrote everything down on his whiteboard at the back of the truck. Next, we set up our incident control point at our truck to interview witnesses. The idea is that you isolate everybody, you bring in one witness at a time and you corroborate their stories, so by the end of it, you have the complete picture. And if you haven't, you go back and start again.

So, at the Clarence, I quizzed the manager. 'Who's in the hotel now?'

'No one, I told you, Captain, we evacuated.'

'Are you sure? Is every room empty? Can you account for everybody?'

The answer was yes. Bearing in mind that our number one priority is preservation of life, we usually go through this process fairly swiftly, because we've blocked the road and wherever we are has come to a standstill. And even though we take our time, we don't just sit there – we have to have a purpose and a plan to ensure that that device doesn't function.

Commandant Boyle wanted to know more about the layout of the cloakrooms. 'How do I get to the downstairs toilet of the Clarence Hotel?' The manager obliged with the information that there were 30 steps down, an immediate turn right, 15 steps down through a swing door, into the second cubicle on the right, the right-hand corner of the cistern.

'That's our robot – will that go down there?' Boyle asked him.

No, he told us, it was too narrow. But before we decided on a manual approach, I realised that what we could do was get the robot into the hotel and to the top of the stairs. While Paddy asked the questions, I drove the robot through the door to the stairs to confirm that there was no way it would go down the steps. However, it could go forward, and I could extend the claw and turn the camera to see the cloakroom door.

Paddy thought for a bit before coming up with his RSP – render-safe procedure. With remote and semi-remote approaches out, he said, 'I'm putting the bomb suit on now, Ray. Pull the robot back a bit. This is what I'm going to do; you brief the team and keep everybody back.' He went down the stairs in his suit, placed the charge designed to disrupt the device, came back upstairs and boom, the bomb was disarmed. We waited for the appropriate soak time, then Paddy got into the suit again and went down to bag the items for forensics. Job over. We were just packing up when an American tourist approached us. 'Oh, my God, you're so brave! We'd like to invite you for dinner tonight.' We didn't go, but it was nice to be appreciated. It had all gone without a hitch, and

I'd learned a lot from watching an experienced pro. I was away on my professional life.

Of course, not all callouts were that straightforward, no matter what training I'd had. Sometimes, events would take on a life of their own and before I knew it, the situation would get out of hand. What was important then was to remember everything I'd been taught and not to panic. Our training has prepared us for any number of situations, even ones which don't go as expected.

One incident that I blush to recall involves a piece of mistaken intelligence that resulted in the destruction of a car belonging to the head of the GAA. In our defence, this happened during the height of the Troubles, when any stray vehicle was seen as a threat. The car had mysteriously appeared one morning in the Curragh, having been stolen in Dublin by persons unknown. There it was, parked by the side of the road on a slight incline. The military police were on it quickly. They tried to open the doors, but they were locked. They reported the car to Intelligence and upon checking the registration plate, it was traced to a former prominent member of the Provisional IRA, Seán MacStiofáin. Alarm bells started ringing, quite obviously, and we were called out to check it out.

As per procedure, we set up our incident control point at the bottom of the hill in our armoured car. We'd just got delivery of a weapons system designed for dealing with car bombs, so, using our robot, we placed devices on the roof of the car and initiated them. They started to burn – no surprises there – through the roof of the car into the interior. However, anything in the car that was flammable started to burn, flames shooting up inside. Again, no surprise; until the brakes burned out and the car began rolling backwards down the hill towards us. *Uh-oh*, I thought, as the ball of fire came hurtling towards us. We would have jumped out – except for the fact that we'd been locked into the armoured car for our own safety!

My heart stopped as the vehicle came barrelling towards us and, as they say, my entire life flashed before my eyes. And then the petrol tank blew up and the car stopped dead, just feet away from us. As we looked on, open-mouthed, it just sat on the road, burning away. *That was close*, I thought.

We knew that it could burn for a few hours yet, so having been released from our vehicle, we gathered together our gear and prepared to return to barracks. The next thing, another car pulled up and out of it clambered a man who identified himself as Seán Ó Síocháin, then Director General of the GAA. He looked at the burning car, mouth open in astonishment. 'That was my car.' Whereupon the intelligence officer disappeared. Obviously, he'd got his wires crossed with the two names, but there was no way he was about to admit it. And as to how the car ended up right outside the Curragh was anyone's guess ...

In fairness, the man was very understanding about it, but he said, 'Lads, my wife gave me a new set of golf clubs a few weeks ago. They're in the boot. Do you think you could retrieve them?' Yours truly went over and opened the boot, retrieving one mangled putter bent in a U shape, which I presented to him in silence.

Looking back, I can laugh, but this incident provided me with a valuable learning, and the work of bomb disposal is all about learning lessons. After each incident, we have a debrief to run through the incident and what we've learned from it, and more importantly, we ensure that we apply this learning later. The above operation is a very good example. Setting up our incident control point at the bottom of a hill? We wouldn't do that again! Locking ourselves into the vehicle? Absolutely not.

———◆———

A short while ago, I happened to drop into Cathal Brugha Barracks, as I sometimes do, and one of the guys asked me if I remembered the time when a shell fell out of a wall in Coolock. After hundreds of callouts over 45 years, I drew a blank. 'Remember, boss, the time they told you you'd killed a small child on a tricycle?'

It all came back to me. I was still a rookie at this stage when I got a callout about an artillery shell that a man had knocked loose while taking down a wall. Apparently, years before, someone had found it in the Glen of Imaal and had brought back to their garden, then had got worried and buried it into a wall. And now it was lying on a suburban pavement, surrounded by the broken remains of a garden wall.

After the usual preparations, I went out and looked at it. I was alarmed. I could see glistening on the fuse, which meant that explosive was leaking out of it. *We don't want that*, I thought, because it could go off at any minute, and in a residential area full of children out playing on the street. The first thing to do was to evacuate the immediate area, which presented me with one problem. A Garda said to me, 'See that house over there? That bedroom window? There's a lady in there who's terminally ill and can't be moved.' It was like an exercise, I thought. Only it was for real.

The obvious solution was to blow *in situ*. It couldn't be moved, but on the other hand, if it detonated, the damage would be enormous. I needed to think a bit creatively. Was there anywhere nearby, a bit of open space? There was a large green area at the edge of the estate which might do – and it was at a safe distance from any houses. Bingo. I requested a Land Rover from the barracks and lots of sandbags. Next, I clambered into my bomb suit, the groin protector, then the massive helmet, feeling its weight on my shoulders. I've heard it described as like wearing two wet blankets sitting in a sauna, which is true, but to me, it's like putting on a suit of armour: I might be sweating and clammy, but I also feel more capable of what I know I have to do. Now, with huge care, I lifted the projectile onto the sandbags in the back of the Land Rover. As slowly as I could, I drove it to the field, feeling every bump and pothole in the road. I'd dug an undercut in the grass in preparation and now I placed the shell into the undercut as gently as I would lift a newborn baby. Sighing with relief when it settled into place, I began to cover it with soil, every spadeful making my heart beat a bit faster. Then I set the charge for the controlled explosion, made sure that the area was completely free of people and blew it. The explosion was massive. I breathed a sigh of pure relief. The situation had been outside the norm, but I'd managed to come up with a creative solution and, more importantly, no one had been injured in the process.

Satisfied with my work, I went back to the barracks, wrote my report on the incident, then went home. I was still living in Rathfarnham and in those days, we did duty from home, so I went home and told Daddy all about it. He was fascinated by my choice of profession and loved to hear the stories about what had happened in my working day.

Then the phone rang. When I answered it, it was an officer on duty in the barracks. 'Captain Lane,' he said crisply. 'Just to let you know that in your operation today, a child was killed by shrapnel.' There was a loud ringing sound in my ears as my panic grew and I was hardly able to listen to the rest of the call. Seemingly, after I'd blown the shell up, shrapnel had travelled over a nearby house; there was a child on a tricycle in the back garden and the shrapnel had hit him and killed him. I went white. I'd killed a child? How on earth had that happened? I went back over every step of the process, trying to work it out. Daddy couldn't believe it.

I sat there, head in my hands, until an hour later, the phone rang and it was the duty officer again. 'Delete killed, substitute nearly killed.' The colour returned to my face. *Thank God*, I thought. I was so relieved. To think I'd killed someone, and a child at that. But then I began to wonder: how had the poor thing been 'nearly killed'? Had a piece of shrapnel really travelled all that distance? I had done everything possible to mitigate any injuries, covering the shell in soil to ensure that it wouldn't splinter. That night, I didn't sleep a wink.

The following day, it got worse. A newspaper got hold of the story and the next day, a photo appeared of a small child on a tricycle in his back garden, a piece of shrapnel on the ground beside him. The headline read: THE CHILD WHO NEARLY DIED. Clearly this story was going to run and run. I went into work that day, head hanging, only to discover that, thankfully, the child was completely uninjured. The piece of shrapnel had been taken home to the back garden as a souvenir, it transpired, and there was no harm done.

The colour returned to my face as the sheer relief set in. I didn't even mind when I earned myself a rap on the knuckles from my boss. 'We won't have metal falling out of the skies of Dublin,' I was told. I was so happy that I hadn't killed the poor child that I agreed. I didn't point out that metal hadn't actually fallen and that I thought I'd done a great job in the circumstances! I knew that I had to pick my battles. With the Troubles in full swing at this stage, we in the Ordnance School were going to find ourselves properly tested.

CHAPTER 6

'YOU'D BETTER LOOK UNDER YOUR CAR'

t was 1986, during the height of the Troubles, when I got a callout to a suspect vehicle, a van, in the car park of the Carrickdale Hotel in Dundalk. It's a very well-known hotel, popular with families, and has great views of the Cooley Peninsula, but at six o'clock on a dark winter's night, I wasn't very interested in the scenery. When we arrived at the scene, the car park was in darkness. There was an eerie silence and it started to rain, so it was absolutely miserable. I greeted the Garda in charge, a good friend of mine, Inspector Mick Staunton. 'Is there anyone in the van?' I asked.

'No,' he replied, to my relief. That would have changed the situation entirely. It doesn't matter who, if there is someone inside a vehicle, our only goal is to preserve his or her life. So, I requested that everyone be evacuated the required distance and I told my sergeant to deploy the robot up to the doors to get a closer look. Then we put a little explosive charge onto the sliding side door of the van to open it. It's called a shape charge, and basically, we cut it to match the shape of our target, in this case the lock on the door. This might seem straightforward, but a lot of practice goes into getting that charge right. If I put a kilo of explosive on a car door, I'll blow the entire vehicle up. We also become familiar with the different charges required on different vehicles: for Japanese cars, for example, the metal is lighter, but more is needed for heavier Swedish vehicles. Using explosives isn't a precise science, but the principle remains the same: there's no earthly reason why you should do any more damage than you need. So I attached my charge to the van's door and eventually it slid open. I got the arm of the robot in and shone a light to see what was in there.

Then I saw a face: two eyes, a nose and a mouth.

'There's somebody in there,' I said to Mick. What's more, where there had been silence, I could now hear the low murmur of conversation. 'Get the bomb suit,' I said to my sergeant. There was nothing else for it but to go and have a look myself. I put on my bomb suit – not a quick process – then walked forward to the opening where the robot was, and I shone my torch into the van. Nothing. What on earth was going on?

I went back to my van and took a closer look at the monitor. *For God's sake*, I thought. We were looking at a plastic bag.

Eventually, we worked out what had happened. The arm of the robot had gone into the van and lifted the plastic bag on its arm, twisting it into a shape that looked like a face. Then, when we'd set off the charge to open the van door, we'd triggered the radio. It was comical, but when I thought of what might have happened, I wasn't laughing. If there was somebody in there when we'd set off the charge, they'd have had a bad headache at the very least! I went back and I had a cup of coffee with Mick and we both managed to have a laugh about it.

In my business, we have a saying that an officer is limited not by what's in his van, but what's in his brain. For example, if I want to penetrate a gas cylinder, but I'm worried that the explosive charge I'm using is too powerful, how do I reduce the strength? I put the explosive charge further away from the target. How? Using a Tetra Pak milk carton: placed on its side, it provides a gap, which reduces the power of the explosive. If you want to mitigate it further, you can fill the carton with sand and that reduces the charge more. It's called innovation, and in bomb disposal, we're learning all the time.

—◆—

The Carrickdale callout was one of the lighter moments of our work during the Troubles. From the 1970s to the 1990s, the IRA was the most sophisticated and best-organised terrorist organisation in the world, so we needed to be at our best to keep up with them, never mind being one step ahead.

By some estimates, 16,500 bombs and 2,000 incendiary devices had functioned prior to the 1994 ceasefire. Because of the sudden massive increase in devices we were no longer humble bomb-disposal officers, but EOD or IEDD officers. We were rostered and on call 24 hours a day in various barracks around the country to deal with endless callouts as the IRA developed ever-more-sophisticated weapons. Along with the thousands of bombs and IEDs, the IRA also developed 15 types of hand grenade, 19 types of timer and power units and, crucially, 22 different mortar systems.

The mortar bomb was something of an IRA speciality. They were constantly developing the mortar's capability in its many forms, from Mark 106s to Mark 16s. In brief, a mortar bomb is a straight piece of hollow metal tubing. At the top is a serrated nut with wings on it. When the mortar is launched from a vehicle, or from a person's shoulder, the wind catches the wings of the nut and rotates them, locking the bomb into the armed position. The tube contains the explosive and a TPU, and a flashbulb is placed into the tailfin unit. So, when the TPU reaches its designated time, the bulb flashes, the incendiary mix catches, the mortar flies through the air, it locks into armed position. As it hits the ground, a floating detonator in the bomb goes into the bottom of the nut, causing a detonation. Boom. It's rumoured that IRA activist Rose Dugdale had a part in developing the mortar bomb.

We had a great deal of professional admiration for the device. Fortunately, we were given a number of safely detonated mortar bombs by the Gardaí, along with a range table which had been made out by the IRA. This allowed us to replicate what they were doing in the safety of Finner Camp in Donegal. The Mark 6 was particularly impressive. The body was nice and smooth so that the bomb would safely leave the tube, but with serrations on the bomb, when it detonated, there would be more fragments. By the time we got to the Mark 10 in the late 1980s, the mortar bomb was capable of being launched remotely and each tube could contain 25kg of explosives.

The Mark 10 consisted of nine tubes of explosives, launched from the back of a truck on an A-frame. That's a lot of explosive. Inside the cab, under a coat, was the TPU. The idea was that the driver was told where to go and a precise spot was marked on the road. They would park, remove the coat to reveal the TPU, activate a number of switches to arm the mortar bombs, then get on a bicycle and cycle away. After a time lapse, the first bomb would leave the tube, then the truck would settle for a few seconds before the next mortar would launch, and so on. There was one weakness: if the electronic component at the base of each tube failed to ignite, the fusing system would still have been activated, so the mortar bombs would detonate in the truck.

Those of us who remember the infamous Downing Street mortar attack in February 1991 were both terrified and professionally impressed. To set a mortar bomb in a vehicle a full mile away from the Prime Minister's residence and fire it so that it lands in the garden of Number 10 was phenomenal. The mortars had been fired through a hole in the roof of an abandoned van, so obviously there'd been considerable surveillance by the IRA on that area. When the IRA's work was done, the van blew itself up. In 1994, the organisation launched a series of mortar bombs at Heathrow Airport, again, fired from a van parked close to the airport perimeter fence. None of them did any damage, but that was deliberate. The message, that the IRA wanted progress in the peace process, had been delivered.

Another famous example of IRA ingenuity came with the sudden disappearance of quite a lot of McVitie's digestive biscuits from the grocery aisles of shops in Northern Ireland. Only McVitie's, by the way … It turned out that the IRA were using them, not to dunk in their tea, but to stop the recoil when they fired improvised grenades from a shoulder launcher. To give you context, we tested a mortar bomb from the arm of a robot, and it broke the arm, such was the strength of the recoil. If this happened to a person, they could lose an arm. They needed to get something that weighed the same as the grenade that they were launching and came upon digestive biscuits. Two packets, wrapped in J Cloths, acted as a counter mass to the projectile when it was fired, and a shower of biscuit crumbs would shoot out of the back. A briefing diagram from this time noted that when launched, it 'simultaneously feeds birds and cleans windows to rear of weapon'. Joking apart, it was lethal.

The IRA were also adept at using modified vans and trucks to carry homemade explosives. I can clearly remember one callout in May 1984, largely because it was resolved by our driver. It was the weekend Ronald Reagan visited his ancestral home in Ballyporeen, Co. Tipperary, and the IRA happened to choose this moment to move a truck full of Mark 10 mortar bombs across the Border, from where they would attack a British observation post.

I was the officer on call at the time, so off I went with my sergeant and driver to the scene. We evacuated the surrounding area as usual, and then I took a good look at the truck and its load, wondering how I was going to get the explosives out of the tubes. I couldn't just remove the tubes: we didn't know how stable the explosive was, so it wasn't safe to lift the things out. Also, we had no idea whether something else was at the bottom of the tube that might cause damage.

I hummed and hawed for a bit before my driver had an idea. He was an experienced mechanic as well as having driven every kind of vehicle. He said, 'You can use the arm of a forklift to get over the tubes and we can then make a line that sits over the top and take each one out separately.' I could see what he was getting at. We'd attach a loop of rope to the arm of the forklift and as we'd lift up the arm, it would tighten on the top of the tube and then, hopefully, it would come out. Of course, it didn't happen as easily or quickly as that. We actually rehearsed for a full day, using a similar kind of gas tube, a forklift and a series of loops to make sure that we could lift the bombs out safely. Eventually, we were able to pull all the bombs out, one after the other. It was an example of ingenuity and teamwork every bit as effective as our enemy's.

Getting to know a particular bomb-maker's signature was all part of the game back then. In EOD, we became adept at recognising the handiwork of an individual and at admiring the quality of the product, even if we were determined to dismantle it. I used to know one of the IRA's top bomb-makers by the sheer quality of his bombs. He was always so precise and meticulous: nothing moved unless it was supposed to move, the wires were tidy and Araldited neatly into position. Whenever I got a callout to one of his signature devices, I'd immediately recognise it.

One day, when the Troubles were long over, I got a phone call from a colleague who was at a callout that was puzzling him. A car had been brought into Coolock Garda Station, with an under-vehicle improvised explosive device – a UVIED – which was a bit of a mess. As ordnance officers, we would share intelligence and knowledge freely and would often attend a scene to offer our take on things if asked. Intrigued, I set off from home, and when we arrived, my colleague showed me the X-ray of the device taken by the robot. He was right – it was a mess, a mass of wires sticking out in all directions. The two of us spent a good hour at it, we filled our board with ideas, we drew diagrams and all that, but it made no sense.

Then the penny dropped. I took out my mobile phone and I rang this guy I knew in Dundalk. A member of the Provisional IRA, he had been a feared adversary during the Troubles. I had a hunch that this device was one of his and that the mess we were looking at had been created deliberately. Rather than saying, 'Did you make it?' I said, 'I'm looking at one of your devices here.'

He replied, 'And you'll note it won't work.'

'I can see that. What's the purpose of it?'

'There's a message there.'

'What's the message?'

'The message is that people like me were left behind in the Good Friday Agreement. Make sure the message is passed on.'

'Hang on, I'm a soldier,' I protested. But he knew that I had no choice but to include what he'd said to me in my report. Apparently, promises had been made to people like him that hadn't been kept, so I passed on the message. In my report on the incident, I said that the device was an expression of his annoyance and that he was a dangerous man to annoy.

◆

Because the IRA was so ingenious, part of my job in the Ordnance School during this period involved technical intelligence gathering. Stationed in Finner Camp in Donegal or in Aiken Barracks in Dundalk, I would often find myself having a quiet pint in a pub in the town, picking up some tidbit from a member of the organisation, or from someone in the know, that might well turn out to be useful. I was never under the impression that my counterparts in the IRA were harmless, but let's just say there was a grudging mutual respect.

I can still remember meeting a senior member of the IRA in a pub somewhere along the Border. I arrived early and I went in, and they wouldn't serve me until this guy at the back clicked his fingers at the barman and said, 'Give him a drink.' I assumed that this must be my contact, but next thing, three Land Rover-type vehicles raced into the car park. The door of the middle vehicle opened and out came my senior IRA member. He and his entourage came into the pub, and we sat down for our chat. Generally, we talked around the houses and then a little nugget of information might be dropped, before the conversation ended. Now, once our chat was over, they all raced out again, but as the senior guy got to the door of the pub, he turned and looked back at me and said, 'You'd want to check under your car tonight. Remember one thing, whatever you can build, I can build it better. In fact,' he added, turning to the other people in the bar, 'you'd all better check under your cars.'

Not wanting to look as if I was spooked by my contact, I finished my pint, watching them race off into the night. Soon after, I waved the barman goodbye and went out into the car park. It was pitch dark out there, and when I turned to look back, I saw the barman standing at the window, looking out to see what I would do. I wasn't stupid. I got my torch from the glove compartment, had a good look under the car, then drove off. I knew that they wouldn't have done anything, but it paid to make sure.

———◆———

Another part of my duties at the time was to wear a 'technical intelligence' hat, which meant that I would liaise with different people and countries to find out what was happening out there in the world, where the threats were, what was coming down the line and so on. The difference between technical and intelligence is really not important. Really, anything to do with a bomb, I wanted to know about it. And of course, part of my job was to pass the intelligence on.

I used to go to MI5 HQ in London every six months to brief them on anything I'd noticed in my area, along with my counterpart in the British Army. Harry (not his real name) and I became great friends. We used to meet for a few drinks the night before we were due in HQ, but when we arrived the following day for our meeting, we'd have to pretend we didn't know each other very well. After all, we were operating on different sides of the Border and the relationship between the British and Irish governments at the time was professional, if not always easy. In the Ordnance Corps, we had to be called out by the Gardaí to incidents; in the North, the RUC would call on the British Army to deal with any devices it might find. That was the way it worked on either side of the Border and the political sensitivities of the time had to be observed. There was to be no fraternising without prior approval and the Border was on no account to be crossed if not permitted at the highest level – or at least, not without good reason.

It was all a game, even if the outcome was deadly serious.

At the time, our liaison in MI5 was a woman and, in the pub, Harry happened to mention to me that she drove a huge motorbike. I said, 'You're joking.' Anyway, we went to the meeting the next day and, of course, it was all very formal. We shared our 'technical' intelligence and when we were finished, I happened to say to her, 'How do you get into London every day?'

'Oh, I have a BMW motorbike, top of the range,' she said. I looked at Harry and he looked at me and he winked. These encounters, and the friendship with Harry, provided a rare bit of light in what was a dark and miserable time in Northern Ireland. I knew that as my counterpart in the North, Harry was under a huge amount of pressure. Unlike me, he was a walking target for republican paramilitaries, so anything we could do to find lightness in the situation, we did.

Before the next briefing, six months later, Bridget said to me, 'Do you know what, I'd like to go over to London with you.' Of course, I had to okay it with the secret service! 'Absolutely,' was the response. We were to have lunch in the House of Commons, as it happened, so Bridget was excited about this, to say the least.

The dining room was magnificent, a long, wood-panelled hall filled with paintings of various parliamentarians through the ages. I was surprised to see a portrait of an Irishman among them, Richard Brinsley Sheridan, a noted humourist and playwright, author of *The School for Scandal* and *The Rivals*. I made a note to ask Harry about it while we were introduced to our dining companions: Harry, the lady from MI5, and two middle-aged men who had been at the meeting. Bridget introduced herself and, being gentlemen, they got up and shook hands, introducing themselves as 'Mr Green' and 'Mr Blue'. Bridget was a bit taken aback but she hid it well.

The wine list arrived, and Harry ordered the most expensive bottle on the list. 'On the Queen,' he joked. As the lunch progressed, we had a second, then a third. As I chatted to Harry and to our MI5 counterpart, I could hear Bridget chatting away to Mr Blue and Mr Green, who were blotto at this stage. 'Oh, so you're on your second marriage? How does family work with your job? Do you find it hard to keep secrets?' and so on. We didn't need spies when we had Bridget.

The hours passed in a slight haze, until Mr Blue suddenly said, 'Oh, God, what time is it?'

I looked at my watch. 'It's a quarter to five.'

'We were meant to be back in the office at two,' he exclaimed, and they hastily made their exit. Six months later, I rang to organise our next meeting and Mr Blue said to me, 'I hope you're bringing your wife with you.'

'No, she won't be travelling,' I replied.

'Oh, that's a pity, we really enjoyed her company,' he said ruefully.

Even though we were counterparts and friends, Harry and I had never spent time in each other's territory, because it wasn't permitted. So, when he suggested that I come to visit him 'up here', as he put it, I said, 'That'll never be a runner.'

'Put the request in.'

So I did, without much hope, but to my astonishment, they gave me permission to go. So, off I went on the train to Northern Ireland – dressed in my civilian clothes, of course. I got off at the wrong station, which meant that I went into the wrong bar to meet him. Confused, I rang him.

'Ray, do you have a drink there?' he asked me.

'Yes,' I replied, a bit baffled.

'Don't even drink it, just walk out and go north on the footpath and we'll collect you.' I did as ordered, and this car appeared out of nowhere and drove towards me. As it drew level, the door swung open, and I was bundled inside. Harry was sitting in the front passenger seat, and I thought I'd pass out with relief. I wasn't entirely sure how he'd found me, but I was glad that he had.

I was taken to British Army HQ in Thiepval Barracks in Lisburn, where I stayed the night. The next day, they took me for a helicopter spin of all the British positions on the Border. I got briefed on every one of them and it was fascinating. The British Army had developed OPs, or observation posts, at each location to observe the surrounding countryside. Of course, the IRA had cottoned onto this and developed bombs that would detonate at that height. Initially, the British went higher still, but the IRA developed a fusing system that would take the devices to that level. So, the British Army decided to put most of its personnel underground, in large bunkers adjacent to the lookout posts. Again, the IRA copped onto that, developing the barrack-buster bomb – a huge gas cylinder, launched remotely, with about 150kg of explosives in it. It wouldn't go through reinforced concrete, but it would have a good crack at it.

As luck would have it in my game, when I got back to Belfast, I got to see at first-hand how the British Army dealt with IEDs, as a car bomb had been planted in a street just off the city centre. They used two robots but, as I watched the operation, they forgot the platform to lower one of the robots down and it fell off the truck. Now, of course that shouldn't have happened, but Harry reminded me, 'Where's the second robot, Ray?' It was in the car. They had two because they worked from the front to the back of the car. I was impressed.

I went back to the Republic full of my trip, but of course, then I had to bring Harry down here to see how we did things. I was determined to put on a good show, to demonstrate that the Irish were keeping pace with our counterparts in the UK. When we met in Cathal Brugha Barracks in Dublin, I had all the kit on display, so I couldn't understand why he looked disappointed until he said, 'Ray, do you not think I see enough robots in Northern Ireland? Can we not go down to your place?' I got it – the pressure he was under was unbelievable. He just wanted a break from it all. So, off we went to Naas and into Kavanagh's pub and there wasn't any more talk of robots or bomb disposal. Instead, we had a great discussion on the history of Kildare with Commandant Finbarr Lambert.

In my career, I've been driven to a great extent by showing that the Irish can be as good, if not better, than our UK counterparts, who at the time were deemed to be the best in the world at EOD. That rivalry had a purpose: thanks to our work on terrorist devices at this time, we were to become one of the most experienced and sophisticated teams in the world of EOD, exporting our expertise to conflict zones such as Lebanon, Afghanistan, Somalia, Bosnia and, at the time of writing, Ukraine. We learned on the job.

CHAPTER 7

FLAGSTAFF, BORDER CROSSING NUMBER TWO

The Troubles in Northern Ireland left their mark on so many of us, but nowhere was this more apparent than the Omagh bombing. It was a terrible irony that it happened four months after the Good Friday Agreement, but it proved that dissident Republicans were still dangerous, and still intent on murder. 'There's a bomb, courthouse Omagh, Main Street, 500 pounds, explosion 30 minutes.' On 15 August 1998, this warning was phoned in to the busy newsroom of a Belfast newspaper. Thirty minutes later, a huge car bomb exploded, killing 31 people, including two unborn babies. To this date, no one has faced criminal charges over the bombing.

In 2024, the then Secretary of State for Northern Ireland, Chris Heaton-Harris, ordered an independent statutory inquiry into what happened that day. This followed a judgement in Belfast that it was 'potentially plausible' that the bombing could have been prevented, if not for gaps in intelligence and mistakes by both the bombers and the police on that day.

One week before the Omagh bombing in August 1998, I was on duty in Dundalk. Sergeant Liam Nolan, an intelligence officer with the 27th Infantry Battalion, came to me in my room and said, 'Sir, there's a British Army operation in the Cooley Peninsula. I think you should go and check it out.' Now, this guy I trusted, because he had a great feel for what was happening on the ground, and he'd introduced me to a lot of people. It transpired that an Irish helicopter had taken photos of a huge crater just in front of a place called Flagstaff, or Border Crossing Number Two, as we called it. It overlooked Narrow Water, where 18 British soldiers had been killed by an IRA bomb on 27 August 1979, the same day as the assassination of Lord Mountbatten at Mullaghmore, Co. Sligo.

I had a feeling my colleague was onto something potentially important. I was going to observe and get a sense of the situation.

Slowly but surely, I got myself to the Border and there, in the middle of nowhere, I found a huge crater in the ground, on the British side. (By that I mean mere feet away from the 'Irish side' of the field, where I was standing.) Working away inside it was an anxious-looking British Army officer. I introduced myself and I told him that I was a bomb-disposal guy and I'd trained with the British Army in Kineton in 1981 and so on, and I happened to remark that the crater looked very interesting.

'Yeah, sir, it is very interesting,' he replied, looking puzzled and a bit stressed.

I pointed to a tangle of poles that lay beside the crater. 'What are these poles here?'

'I don't know, I'm trying to work it out.'

I had an idea. 'If you stand those poles up, you'll find holes close by. Put them in the holes, stand them up and let's measure the distance between the poles. I think you'll find they're equidistant.' This I did without crossing the Border, needless to say.

He did as instructed, lifting each pole and placing it carefully into position. Bingo. 'Oh, they were testing explosives,' he said.

'That's right,' I said. 'There are also serious burn marks around here, aren't there? That's unusual because when you get detonation,

82

it generates a powerful shock wave, which pushes things back, so it's unusual to have a flashback go so far.'

'Hmm,' he agreed. 'I wonder what that is?'

'Well, they could be testing some sort of flammable addition to the explosive,' I said. 'Hang on. I'll be back in a second.' I went hunting on the 26-county side of the Border, where I found the remains of a gas cylinder. The explosives had clearly been housed in it, which would account for the burn marks. I brought it back and said to him, 'Now, everything I've told you, make sure it goes into the report. This has to be reported by you, not me.' As ordnance officers it was part of our job to keep abreast of any developments without compromising security or causing any trouble. What I'd seen on the Cooley Peninsula made me feel uneasy, and I followed my instincts, and Liam's intelligence, to see what was happening on the ground.

Ironically, I was at the Giant's Causeway with my family on the day of the Omagh bombing, and I said to Bridget, 'I'm going to go to Omagh, I'm going to call in. I have to.' Her reply was 'Not a chance. You're not going to put your family at risk by going.' She had a point, but I was raging that I didn't go. Of course the bomb had already been detonated, but I couldn't help wondering about the forensic information that would remain – concerns that I later discovered were entirely valid.

Some time later, I happened to be listening to a news report on the bombing, which mentioned the terrible burns suffered by a lot of the injured. That encounter on the Border suddenly made sense. Should I say something, I wondered? Would it make things any better for those left behind, or would it add anything to the understanding of what had happened that day? I had no idea, but I felt that I had to do something.

With the memory of what had happened out on the Cooley Peninsula fresh in my mind, I got hold of the telephone number of the solicitor representing the families of the deceased. I introduced myself and told him I had a story. 'It might be applicable to what happened in Omagh,' I finished.

There was a long silence. 'Why do you say that?' he said eventually.

'In everything I read [about the bomb], people suffered severe burns, and with a detonation, it's unusual for people to suffer in that way ... that's what I'm reading, and maybe what I'm reading is wrong ...' my voice trailed off.

'Keep going.' There was a stunned silence as I told him about my encounter with the captain at the bomb crater. I finished with 'Can I ask you a question? Have you looked at all the British Army reports for all the incidents prior to Omagh, say three months before Omagh?'

'We have.'

'Well, therefore you have to have this report.'

They did have reports, he told me, but details had been redacted. 'But of course now, we have the public inquiry coming up here and everything has to be on the table. There can be no redactions. We'll get everything.'

I sincerely hoped that they would. There had been many reports about intelligence gathering and phone calls that may or may not have been passed onto the RUC, but I also knew that the bombers themselves had given inaccurate instructions to the media and to the Samaritans in two phone calls, instructions which led to people being funnelled towards, rather than away from, the bomb. Whatever the circumstances, I felt duty bound to tell the solicitor what I knew. 'I found the remnants of a gas cylinder. I gave it to the British Army,' I told him. Then I said, 'I'm going to ask you a question. If you can't answer it, don't answer it. What was the bomb in Omagh contained in?'

I knew the answer. A gas cylinder. Now, if you take this to its logical conclusion ... there was an incident in Cooley and serious intelligence to be got from it. Somebody either deliberately, or stupidly, decided not to act on the intelligence they had that day. The dissidents were out there testing explosives for a purpose. And it wasn't just Omagh. There was a bomb in Banbridge around the same time.

The other thing that concerned me was forensic evidence. The solicitor seemed to be saying that he hadn't got much, which, quite frankly, surprised me. 'Look,' I said to him, 'this is key. Twenty years before Omagh, in 1974, you had the bombs in Dublin and nobody had any idea about forensic analysis or anything, so the fire brigade came in and washed the streets, so any forensic evidence that was there was washed down, because we just didn't know. The British Army and the RUC developed this significant capability in acquiring forensic evidence over the years, which would have been available on the day of the Omagh bombing. So, if somebody says to you, "Ah well, our procedures were very poor," that's absolute bullshit. Their procedures were top notch, superb.' I knew that the scene in Omagh was isolated very quickly for forensics, so something just didn't add up. I'm not suggesting a cover-up here, but that there might be questions about intelligence, whether its significance was realised and whether it was acted on. Twenty-five years after the Omagh bombing, with the dissident Republicans who were eventually found liable now dead, the families of those who died in Omagh have yet to receive meaningful answers.

CHAPTER 8

HILL 880

Lebanon, 1988

Every soldier looks forward to his or her first foreign tour with a mixture of excitement and trepidation. I was no different. For a career soldier, a foreign posting is often the defining moment of their working life, and it gives them the chance to use their training and expertise in the field.

In 1988, I went to Lebanon for the first time with the 64th Infantry Battalion. I was leaving Bridget and my two children, David, then aged six, and Clare, two, behind for seven long months to go to a country that I'd only ever seen in news reports and which, at the time, was in chaos. At the same time, I'd been told by countless colleagues that the camaraderie and sense of teamwork overseas was unbelievable. Both things were true. And yet there was a third element to my service that I hadn't fully foreseen: that I would see and hear things that didn't seem right to me, and that dealing with them would turn out to be unsatisfactory, to put it mildly. Sadly, also, they would have far-reaching and deadly consequences.

I have always thought Lebanon to be a fantastic place, a beautiful country that has been riven by various conflicts. Run by the French up until the end of the Second World War, it was then occupied by Syria, then Israel invaded after a massacre of its citizens by Fatah, a branch of the Palestine Liberation Organisation. Many of you will remember the vicious civil war that erupted there in the 1980s, with various factions controlling various parts of the country and the Hezbollah kidnappings of American and European citizens including Terry Waite and Brian Keenan.

UNIFIL – the United Nations Interim Force in Lebanon – came into being once the Israelis had withdrawn and a buffer zone was created in the south of the country, of which the Irish formed a part. I think that there's sometimes been a perception that our UN peacekeeping missions aren't that perilous. Far from it. Infamously, 26 soldiers died in the Congo in 1960, and 47 Irish soldiers have been killed in Lebanon. In 1980, the Irish position at Hill 880 was attacked by the South Lebanese Army and three Irish soldiers were kidnapped, with two killed. In 1986, Lieutenant Aonghus Murphy was the victim of an IED, laid by terrorist Jawad Khasfi. Khasfi's name will be important.

And in 1988, during my first posting, three Irish soldiers were killed, also by an IED.

I was a captain in 1988, just 35, but I'd been promoted to acting commandant for my trip to Lebanon, as part of the 64th Infantry Battalion. In an overseas battalion there could be 500–600 men, with all the structures that go with that: there's a battalion commander, an operations officer, information officer and so on. The area in which we work is known as the AO, or area of operations. In that area, you'll have A Company, B Company, C Company – each would be a hundred or so men with a commandant in charge, and their job is to oversee that area, based on the mandate that you have from the UN. What sounds relatively straightforward, however, becomes a great deal more difficult when you are surrounded by hostile forces, caught between the Israel Defense Forces (IDF) on the one hand and their supporters, the South Lebanese Army, on the other. Colonel

Michael Wright, in his monthly report back to Ireland in 1986, had stated that 'Irishbatt troops are now at greater risk than before.' In fact, in an article on this tour in *The Examiner*, Conor Ryan said 'the 591 members of the Irish Infantry Battalion were on as dangerous a tour as any in our proud peacekeeping history'. I would agree.

My job in this posting was solely to look after weapons and ammunition, i.e. one half of my job. There was to be no bomb disposal, or EOD, which will become important later, because responsibility for that area lay with the French Engineer Company. (For Irish Ordnance Corps EOD personnel, theatre-specific training is undertaken before you can perform the duties of an EOD officer overseas.) At an intelligence briefing held before we left Ireland, I got the Battalion Operations officer, Jim Mortell, to ask, 'Are there any IEDs that we need be concerned about in Lebanon?' The answer from Intelligence was a definitive 'no'. However, as it transpired, IEDs had been a threat to UNIFIL forces for some time. In fact, apart from poor Aonghus Murphy, the 63rd battalion had discovered two IEDs in their AO in May, just a few short months before we deployed. Again, these were traced to Jawad Khasfi. I was aware of this, so that was why I'd asked the question.

This was the background to my first tour, but I looked forward to it, and to working on the other part of my job. I'd had more than enough experience with bombs and IEDs at this point so the opportunity to use another skill was one I relished. I had no idea that the tour would be remembered in part for the tragic deaths of three Irish peacekeepers.

On 17 November 1988, a patrol was heading to the Irish position on Hill 880. This was on top of a hill directly overlooking the village of Haddatha. As they drove along, they noticed a strange object buried in a wall. A decision was made to ask the French engineers to come with their equipment and their robots to deal with it, as per their remit. However, it transpired that the French had gone home and there was nobody qualified in the whole of the UNIFIL structure to deal with it.

So I was called in by the battalion commander and I was tasked to do the job, even though I had no equipment. I wasn't happy about it at all. Apart from the fact that I lacked the specialist training for Lebanon, I absolutely didn't want to put my life or my colleagues' lives at risk when we were not equipped to do so. We had no bomb suits and no robots because UNIFIL had shipped the French EOD kit to storage in Italy, no nothing. In fairness, my boss had asked the Swedes, also part of UNIFIL and who had some capability, but their answer was a straight 'no'. I don't blame them. To put a little context on this, in EOD at the time, we were used to operating at a certain 'operational tempo', as we call it. There was a clear structure to our work, from being briefed by the Gardaí, to setting up our incident control point, to requesting backup from the fire service, to ensuring that the area in which we worked was correctly identified and isolated as per procedures. This enabled us to control the incident completely and to minimise casualties. In Lebanon, the ICP wasn't secure, because we had no idea what the area contained in terms of devices, we had no backup, no equipment and, furthermore, we had no idea precisely where the threat was coming from. Yes, we were dealing with 'a device', but how many others might there be?

Reluctantly, I agreed to do the job. The night before we set out, I got my team together – remember, I had no bomb-disposal people with me, so I'd simply selected some reliable colleagues for the job. We planned our strategy on a blackboard. The object was buried into the wall on a rough track up which our vehicles would go on patrol. It was a large device full of nails and packed with explosives, pointing out onto the track, so you can imagine the damage it would have caused had it been detonated. Our plan was to clear a route to the device not with a robot, or anything sophisticated along those lines – instead, we would be using cardboard boxes to lie on and a few bits and pieces of rope to get to the device and to pull it out remotely.

So, with this rudimentary setup, I got to the device, took it out of the wall and examined it, then looked at the whole scene. Something didn't feel right. I climbed safely through the gap in the wall, and on the ground in front of me lay a roll of white tape. For a second, I thought it was the same white tape that I had been using and that I'd dropped it, but as I bent down to pick the roll up, I felt my own roll of insulating tape in my pocket. I swore. Standing completely still, I looked down and noticed a square cut out of the sod just in front of my foot. This concealed a pressure plate. I could feel my blood thumping in my ears as I took it in. If I'd stood on it, it would have blown my leg off. Then the penny dropped: I had been taken to meet Khasfi before the task (he was not yet in custody) and I'd asked him how he'd planted the bombs. He'd lied to my face, telling me that there was no secondary device. And yet, here it was. (It was later suggested that I might hand the components of the IED back to Khasfi!)

Cursing Khasfi's sheer duplicity, I did my job then, setting enough of a charge to blow up the first and secondary devices, then returned to camp to write up my report. I was clear that the device was designed to kill Irish personnel and that any soldiers who weren't killed by the explosive buried in the wall would have been killed by the IED when they stepped through the gap in the wall. In my report, I drew the following conclusions:

1 The quantity of explosives used and the positioning of the devices presented a great danger to life. Any delay would allow the terrorists to reset the devices.

2 The lack of IED cover, particularly in the Irish AO, was 'a cause for great concern'. We needed trained and fully equipped teams to disable them and the fact that we didn't have any placed me in a difficult position.

3 I had asked about the position on IEDs before I'd left Ireland and had been referred to UNIFIL SOPs, or standard operational procedures.

When I shared my report with senior officers in the battalion, a robust debate ensued. Some disagreed with my conclusions that the devices were designed for the Irish, arguing instead that they were meant for the Israelis. I disagreed because of the device's location, which was too far down for the Israeli forces or the South Lebanese Army to venture. I felt so strongly about it that I wrote a letter to the battalion commander on 21 November. In my letter, I clearly stated my view, based on my experience, that this IED was targeted at Irish personnel and that we needed to take proper measures to ensure that nothing happened in the future. There was a top-level meeting held of all the senior people, which I attended, and my opinion wasn't accepted in spite of support from Jim Mortell and others. Jim said, 'Commandant Lane is the expert in this area.' But the ultimate decision wasn't ours. I looked at the boss and I said, 'Sir, you're the boss. I'm telling you. I was there.'

On 15 December 1988, Khasfi was abducted by the Israelis and the place went ballistic. Because he'd been taken out in our area of operations, some of his comrades thought that we'd somehow been part of it. Khasfi had killed Aonghus Murphy, so the conclusion was that we might well have allowed it to happen. You could have cut the tension in the Irish camp with a knife. As our boss, Colonel McMahon, told HQ in Dublin, 'Situation tense. IRISHBATT on full alert.' In fact, in retaliation, Khasfi's group kidnapped three Irish soldiers. During the negotiations for their release, a sinister message was passed to the Irish that 'a lot of people want to fish in the dark waters'.

On 21 March, three Irish soldiers, Fintan Heneghan, Mannix Armstrong and Thomas Walsh, set off up a track towards an abandoned Irish post to collect rocks to fortify the Irish positions. Concerns had been raised about this practice, because of the possibility of landmines, but it continued. There was a massive explosion and the three men were killed. In Frank Callanan's report, we learn that 'the explosion was audible over a wide area and was visible from the blockhouse of Post 6-9B. Private Michael McDonnell, who was on duty as a sentry in the blockhouse, heard a bang and saw a big cloud of smoke with the truck in the air and saw the truck coming to the ground followed by showers of debris.'

When news got back to us all at Camp Shamrock, we were absolutely devastated. The following day, I went out to the site of the explosion, a huge crater in the ground. By this stage, we had located two robots in crates in UN HQ in Naqoura, so I sent my robot further up the track and I found a secondary device. It was clear to me that the idea behind the two devices was that the truck would go all the way up, and the first device would take it out, and the rescue team would come out, rush into the area, and be taken out by the second one. Instead, because there were only three men in the truck, they hadn't driven all the way up. It was the same MO as the device on Hill 880; the idea there was that when explosives embedded with nails that were placed in a paint tin detonated, they would destroy the vehicle, and when the survivors went through the gap in the wall, they would step on the pressure plate. Writing it now, with the distance of more than 30 years, it seems unbelievable.

When I came back from the bomb site, I went into my room, unable to talk to anyone. Jim Mortell came in with a bottle of whiskey and two glasses and he said, 'Now, let's settle here for a while.' We went through the whole thing and he said, 'You know, Lane, you were right all along. I knew you were right.' I knew that. I didn't need him to tell me, and it didn't matter anyway. It wasn't about me being right – it was about being listened to and having my expertise respected. I'm a professional and I knew how to do my job, I was well trained and capable of assessing a threat, but others hadn't seen it that way.

Jim supported me, a hundred per cent. He was a commandant of many years' standing and he was coming home to be promoted to lieutenant colonel. I was a commandant simply for the duration of the tour, coming home to return to the rank of captain, even though the two of us were the same rank out there. He would often say to me, 'Jesus, man, I can't wait to get you back in Ireland. I can't wait.'

'Why?'

'Because I'll be a lieutenant colonel and you'll be a captain, there'll be none of this cheek,' he'd say. He was an absolutely fantastic guy, one of the best officers I've ever known. In addition I received great support from Commandant Jim Saunderson, the battalion logistics officer. I couldn't have done it without them.

———◆———

I'm going to jump all the way up to 2010 now, more than 20 years after the incident. A report had been carried out by the military, concluding that the events of that morning in 1988 were a tragic accident. I took no further action at this time, returning to my work in the Ordnance School and getting on with my life. I genuinely believed that the families of those soldiers had been looked after in some way. Then, one morning in 2010, my wife said to me, 'Do you know, Ray, I don't know how you can live with yourself. Do you know one of those women is still fighting that court case?' I had no idea that the widow of Mannix Armstrong was still fighting the Defence Forces for compensation for the death of her husband. It wasn't about the money, of course. She simply wanted to know what had happened. Twenty years after his death, she knew that she hadn't received the full story.

The following day I got a phone call from the senior legal officer for the Defence Forces. 'Ray, Mrs Armstrong is still fighting that case and we'd like you to come up and give evidence. You're the last remaining one of the battalion, they've all retired, and they're going to have a pre-trial hearing.' The following day, I went to the Four Courts and they were all sitting there, experts from the Department of Defence and the military.

There was a long silence in the room while I opened my file and I took out my papers. Then I looked at them and I said, 'Please put me in the box and I'll prove that we were negligent.' One of the senior people got up and left the room. Apparently, he rang Mrs Armstrong's legal team and said that they would no longer be contesting the case. Alan Shatter was Minister for Justice at the time and authorised the Callanan report into the death of the three men. Barrister Frank Callanan came down to me in the Ordnance School in the Curragh for a week and I restaged the attack in detail for him. Four months later, his report was issued. This is what he had to say:

> In relation to the five specific matters I was required by the terms of reference to enquire into as part of what was a general enquiry into all relevant matters leading to or surrounding the deaths of Cpl Heneghan, Pte Armstrong and Pte Walsh, the following are my answers:
>
> 1 Whether the standard operational procedures of 64th battalion were appropriate and adequate to ensure the safety of members of the unit, given the operational situation in which the Bn was deployed. **No.**
> 2 Whether adequate cognisance was given to the nature of the operational threats faced by the members of the unit … **No. There was a failure to**

carry out adequate threat assessment and to adopt appropriate force-protection measures arising from that threat assessment.

3 Whether the location where the deaths occurred was 'out of bounds' to Irish troops and if not, whether it should have been so placed. **No. The location was not out of bounds. There was no reason to place it out of bounds ... The route to the Green Rooms should not, however, have been used until after it had been cleared for IEDs and mines.**

4 Whether the device that killed the three should have or could have been detected before it detonated. **Yes.**

5 Whether the persons deployed had adequate training in the circumstances. **No.**

On the day the Callanan report was published, in 2011, a very senior officer in the army rang me on my mobile and said, 'Ray, take a day's leave. This report's coming out.'

I said, 'Yeah, I don't care. I know what's in it and it's spot on.' I'd gone up to Hill 880 with no equipment. I'd rendered safe a sophisticated device. I'd looked at the whole scene as it was and said, 'This is for the Irish.' I did my job. I have no question about it. Other people made the decision that it wasn't for the Irish and the result of that was that we lost three guys in circumstances that, as Frank Callanan stated, were unsafe. To me, it's not about settling scores, or proving that I was right and they were wrong. It's about the deaths of three Irish soldiers because the correct structures were not in place to protect them.

You might well ask yourselves what good came out of the whole thing. Well, UNIFIL definitely looked at the whole area of IEDs and began to take them a great deal more seriously. At the end of my tour of duty, they replaced my role with two officers: one officer with responsibility for ammunition and weapons and one officer with sole responsibility for IED disposal, or IEDD – and they sent out all the sophisticated equipment from Ireland that they'd need. Unfortunately, however, in spite of the deaths of the three soldiers, the lessons didn't endure – as I would discover on my next tour of Lebanon, nine years later.

As you'll see in the next chapter, my role in 1997 did not involve bombs of any description, which came as something of a relief to me. I was the food officer, stationed on the Mediterranean coast, happily looking after eggs, fruit and other things, when a senior Irish officer called me in and asked me to look into a problem they were having out at the Irish Battalion AO. The IDF had mined the route that led to an Irish observation position, which obviously meant that Irish personnel could neither get in nor get out of this position without being blown to pieces.

I drove up the hills to the Irish Battalion, remembering what had happened nearly ten years before. When I arrived up at the 81st battalion, I was briefed by the commander. We had a long chat, and as he outlined the situation, I became increasingly uneasy. Apparently, after the IDF had mined the route, the Irish had – rightly – complained to UNIFIL headquarters and the Israelis had agreed to go back with a JCB and clear the area, which they did. However, the Irish were unhappy that they had no quality assurance, if you like, of what the Israelis had done. They were worried about the possibility of a stray mine being left, so an Irish Battalion search team and an IEDD team had gone in to search and destroy up to one metre left and right of the route. Unfortunately, one soldier had stood on an anti-personnel mine and had lost his leg in the process. I was now being asked to go in and remove anything left behind.

I was shocked. I found it hard to believe that a search team and an IEDD team had been working in the same area. If you think about it, the search team searches for the devices, then withdraws, and the IEDD team goes in to remove them. If they are both working in the same space, it's an accident waiting to happen: more people can be injured and you can't control the situation.

I had no EOD licence and no team, but because the request came from my own country, I reluctantly agreed. As I hunted around for suitable personnel to help me, I had a creeping sense of déjà-vu. I had one guy with me, a captain, who was very level headed. I said to him, 'If you see me doing something in the minefield that doesn't look right, or if you think I'm overextending myself, you just say it. You tell me to come out and I'll come out instantly, I won't argue with you.'

The captain agreed to stand at the edge of the minefield with binoculars, watching every step I took on the rocky path. I had worked out the gap between my steps and had placed enough explosives in the ground to ensure that there were no mines, then I'd blown a hole in the ground to clear it. I repeated the process over and over again until I could walk safely along the path. This might seem straightforward but try doing so in a minefield in which someone has already lost a leg, in 38-degree heat, without any technical assistance.

That was one day's work, after which I returned to my HQ down in Naqoura. The following morning, at the briefing, the UNIFIL operations officer happened to mention that a Commandant Lane had been involved in the operation the previous day. My boss, who was Polish, looked at me and said, 'I didn't know there were two Commandant Lanes in UNIFIL.'

I said, 'Sir, that was me.' Whereupon he went completely ballistic. 'That's it,' he fumed. 'You're not going up there.' I could see his point of view. I was his food officer, so in his mind, I had no place being up there. However, in this case, my Irishness trumped my role as food officer, and anyway, I told myself, my Irish commander had told me to go up there – or he was about to.

On day two, I went back up, I continued my work up and down, removed all the kit that had been left behind, did a detailed analysis of the minefield and compared it to the briefing I'd got from some of the people injured in the original explosion. And this is important: I was told by one of the people who was involved in the incident that they had seen improvised explosives on the left and right of the minefield. I found none. I wrote a detailed report on the incident, and summarised by saying how surprised I was to see the search team and the IEDD team operating together in this situation. What had happened in the minefield was unsafe and contravened every regulation. Frankly, I was surprised to see that after everything that had happened in 1988, this kind of mistake was still happening.

CHAPTER 9

BARBARIANS

Bosnia, 1992

'Out here, Mr Lane, we do that [mimes pushing soil away with his hand] and we're barbarians. In your country, you have to dig a little bit deeper.' These were the words of Valentin Ćorić, leader of the Croatian military police. In that small gesture, he conveyed the essence of the war in Bosnia to me: all it took was the removal of one very thin layer of civilisation to reveal the barbarity underneath. His determination to cleanse his country of Muslims resulted in thousands of detentions and ultimately, in a 16-year prison sentence for him. He wasn't the only war criminal produced by this savage war, but he was someone who symbolised 'the banality of evil', as Hannah Arendt's phrase puts it. I knew him well: I'd spoken to him often and had got to know him during my posting to Bosnia. I'd met his family and friends, yet at the same time, I knew that he was guilty of the most despicable crimes.

If you were around in 1992, perhaps you might remember the conflict in the former Yugoslavia. The Siege of Sarajevo, Sniper's Alley and the destruction of the bridge in Mostar … the whole of the former Yugoslavia erupted into war, as the six federations that had formed the country began to declare independence. There had been rumblings for years, but the depth and speed with which the violence exploded took many of us by surprise. It became the biggest outbreak of mass violence since the Second World War, a bloody civil war marked by ethnic cleansing and genocide. An estimated total of 140,000 people were killed and a further 400,000 displaced. The war seemed all the more shocking because it had broken out in Europe. It involved fighting on streets not that different to our own. The very worst atrocities were to take place in Bosnia-Herzegovina, home to a substantial Muslim population known as the Bosniaks, but which ethnic Serbs and Croats wanted to claim for themselves.

In 1992, I was nominated to go to Bosnia-Herzegovina as part of the ECMM, the European Community Monitoring Mission. The mission was both a military and non-military one. In fact, the EU was making a cautious re-entry into the monitoring role, because two Belgian monitors had been shot dead 18 months before, so we were well aware of the dangers.

My role would involve no bomb disposal. I was actually going as a diplomat, with two people in my immediate chain of command: the Portuguese ambassador, José Pires Cutileiro, who was heading up the mission, and a British general, General Cranston. I was unarmed, had to wear white and worked with serving and retired personnel from Europe and Canada.

According to Lt Colonel Remi Landry: 'The Brioni Agreement, signed by all the parties in the dispute, legally established the mission for the first time, initially as a monitoring mission with the aim of easing the withdrawal of the Yugoslav Army from Slovenia.' The ECMM had been instituted in 1991 and this description by Landry sums their aims up beautifully:

> The initial role of these teams was to monitor the implementation of cease-fires and the withdrawal

of troops. As the overall situation evolved and as the mission became more attuned to the process, the teams ended up doing much more than that. As of 1992, the ECMM was already active in preventive diplomacy and in confidence-building actions. For example, the Joint Commissions, created and chaired by ECMM staff to oversee the implementation of cease fire agreements between the warring parties, proved to be one of the most efficient peace-building confidence measure [sic] at the time. Other confidence-building measures, including humanitarian tasks, were undertaken by the ECMM staff such as: the passing on of messages and letters; installation of hot lines between the warring parties; delivery of food and other humanitarian supplies; mediation tasks during hostage crises and civil unrest incidents; visits to prisons; exchange of war prisoners; and humanitarian convoys.

That was my job in a nutshell. It sounds relatively straightforward, and yet it took its toll on me like no other conflict I've witnessed. Bosnia and what I saw there changed me for ever.

It's funny to think that Split is now a town favoured by students for cheap holidays, and the Hotel Split, the EU's HQ during the war, is now described as a beachfront paradise. When I arrived in 1992, it was very much on a war footing, bustling with civilian and military personnel and with dazed elderly refugees from the war. My first task, after meeting my bosses, was to go to the Republic of Serbian Krajina, north of Croatia. It was an enclave of 200,000 people, ethnic Serbs, who had declared independence. (This had actually led to the Croatian war in 1991.) I was sent up there to repatriate people in and out of the enclave. There would be families all over Europe whose loved ones were in Krajina, so I would get them through the Croatian border, help them to visit their families, collect them and bring them back again. I was also looking after people and giving them food. The Serbian people in Krajina were lovely but they knew their days were numbered, that's for sure.

It's amazing what comes back to you, but I can clearly remember one exchange that really brought home the madness of what was going on. A Serbian lady who had been married to a Croat had split up with him at the outbreak of war. He had run off with their child to Croatia, so it was our job to go down to his home town to retrieve the child – legally, I must stress – and return the baby to the mother. The husband came with us to the border and the scene has stayed with me ever since. There was a white line in the middle of the road, and she was one side of the white line, and we were on the other side with this guy, holding the baby. They were both in tears. I told him, 'You have to hand the baby over now.' They obviously loved each other, they had this beautiful child, and they were both going to turn around and go in different directions. That's exactly what they did, she with the baby and he alone. I was in the middle, standing at the white line, looking at that young couple and thinking, *my God, is the war doing this? What on earth is it all about?* My experience along the Border had given me plenty of insights into sectarian violence, but I'd never witnessed anything like this.

It would get far, far worse.

Out of the blue, after about three weeks in Serbian Krajina, I was summoned by the Portuguese ambassador to the HQ in Split. When I arrived, the place was frantically busy, with people coming and going, getting ready to re-enter Bosnia, preparing logistics and that kind of thing. I had been part of a recce team that had gone into Bosnia before the full deployment of the monitoring team, but it was cursory, to put it mildly, as official after official was presented to us, none of whom I could remember. However, now here I was in front of the ambassador.

'Mr Lane,' he said, 'we've been looking at you for the last while and your performance has been exemplary.' I was astonished. My 'performance' thus far had consisted of the aforementioned repatriations and giving out loaves of bread from the back of a truck. 'So,' the ambassador informed me, 'we've decided to put you in charge of the mission going back into Bosnia-Herzegovina.' I was to be the fall guy – I knew it immediately. Sure, I was European, from a neutral country, and that was important, but I also knew that whatever went wrong in the ECMM mission in Bosnia would be on me.

Nonetheless, I did as I was told. I spent four days putting the team together to enter Bosnia. I prepared orders, a description of our tasks and logistics, which were complicated. I'd never organised such a complex mission in a war zone. We had to plan accommodation, feeding, fuel, maintenance of vehicles and communications – we had to be as self-sufficient as possible. However, due to the significant threats, our force-protection plans had to be of the highest standard to ensure the safety of the monitors.

Putting together the right team is the number one priority in situations like this, so I'm always on the lookout for promising people. Over the four days of planning, a Danish captain caught my attention. Now, he was wild-looking, as if he'd spent too long in the open! However, I listened to him in the hotel one evening and I liked what I heard. He seemed to have a good effect on people around him and a knack for finding the right people to work with. His name was Volmer Svendsen. He really impressed me, so I offered him the job of being my deputy.

He jumped at the chance, but it came at a price. He said, 'You know that I have a reputation for being a bit wild.'

'You look a bit wild,' I agreed. 'But just so we're clear, this is a no-alcohol mission. Do you understand that?

'Yes, Mr Lane,' he said, solemnly.

I wasn't entirely convinced.

The following day, we drove down the Dalmatian coast on a road that wound between the mountains and the Adriatic Sea, past places that had once been holiday resorts, like Podgora, where I'd stayed with my family a few years earlier. It was hard to believe that the pockmarked buildings and burned-out barns had once been intact. On the border between Bosnia and Croatia, we ended up in a little town called Metković, near Medjugorje, the site of Catholic pilgrimage, Volmer was in the lead vehicle, and he pulled in and got out of the car, heading towards a small supermarket. *What's he up to?* I thought, waiting impatiently for him to emerge. After a few minutes, out he came carrying two crates of Carlsberg.

I got out of the car and eyed the crates. I said, 'We're not starting well here, Volmer, are we?'

'Mr Lane,' he said. 'Where we're going, some of the young people [monitors] we have here won't last five days. If I see people I think are in trouble, I'll bring them into my room and I'll give them a few beers and I'll get to the bottom of it and then I can decide, do they stay or do they go. I won't drink any of it myself.'

What I would later learn was that Volmer had unbelievable emotional intelligence. Due to the stress of the job, we had a change of monitors every three weeks and there were young guys (all the monitors in Bosnia at that time were men) coming back in tears about what they'd seen. I'd look at him and he'd look at me and he'd usher them into his room, and I'd hear the sound of the cap coming off the beer. He'd then come out to me and he'd say either 'He has to go back,' or 'We'll keep him.' He kept on top of it. It might have been unorthodox, but it was brilliant as a human resources policy. Volmer was a born leader and an inspirational man.

The cursory recces we'd done hadn't identified anywhere where we could stay, so we arrived in Bosnia looking for accommodation, would you believe, for 35 people and assorted vehicles, radios, communications, everything you can think of. Eventually, we arrived in a little town in the valley near Mostar, known as Široki Brijeg. We found a large, unoccupied house in the town and we all moved in there. Over the next few days, we travelled around, scouting for different locations for my teams to monitor the situation on the ground. I was to meet all the key leaders, military, civilian and religious, and to develop an overall view of the situation to report back to the ambassador. Apart from that, I could interpret my brief as loosely as I liked.

One of my first tasks was to get a local translator to help us to understand Serbo-Croat. In Široki Brijeg, I heard a girl speaking English, and it turned out that she was Australian, so I hired her on the spot. Her family were originally Croatian and she was visiting them, she told me (a lot of people with ties to the country had returned to fight). While I was happy to get her because she spoke fluent Serbo-Croat and English, I knew that she was reporting on everything we said and did to the Croatian leadership, known as the HVO. There was very little I could do about this, because my translators had been recommended to me by the local government. Still, suffice to say that I became wary and vowed to improve my Serbo-Croat to ensure that I was getting the correct translations. In all my time in Bosnia, I only had one Muslim interpreter.

One particular night, all the windows were shot out, and our car tyres were frequently punctured. I used to tell my staff, 'Whatever people tell you, you disregard 99.999 per cent of what they're telling you and of the remaining 0000.1 per cent, you disregard 99.999 per cent of that.' These people weren't intrinsically dishonest – with the country split into warring factions, it was rife with suspicion, rumour and downright lies. A sense of menace hung in the air and the atmosphere was poisonous.

Nonetheless, we soon established a routine: every Friday I'd have a conference and invite the individual team leaders who were stationed all over Bosnia to report to me, and eventually we'd get an overall picture of what was going on. Then, on Saturday, I'd go to Split and report to the ambassador. With the local people being very suspicious, I felt that it was important to make clear that we were here to help. I went to every christening, wedding and funeral,

My wonderful parents,
Ted and Eileen Lane.

The official opening of Defence Forces university accommodation
in Galway by President Childers in 1973. I am sporting a bandaged
forehead after an 'argument' with a weapon!

Appalling Dublin bombings on Talbot Street, 1974.
(Associated Press / Alamy Stock Photo)

Christmas in Lebanon.
Santa's present was an improvised explosive device (IED)!

With the much admired and highly respected Comdt Jim Mortell in Lebanon.

Dressing up for medal parade in Tibnin, Lebanon. 64 Irish battalion.

Arriving at ECMM HQ in Split from Bosnia Herzegovina
for the Ambassador's weekly conference.

Destruction of a bridge in Mostar as the
beautiful turquoise Neretva flows by.

Stari Most bridge in Mostar – a beautiful structure that facilitated the meeting of east and west cultures. All life moved over this bridge. We spent a lot of time monitoring it, as we were aware that elements wanted it destroyed. Sadly, the bridge was destroyed in November 1993 by the HVO.

Posušje camp, Herzegovina, and the dreadful plight of these poor 'residents'.

A joyful but solemn occasion as we reunited
a family in Knin, Krajina.

The Irish desert-pattern
uniform blends into the
Afghan landscape.

Taliban 107mm rockets found in an explosive cache. *(Afghanistan Archive / Alamy Stock Photo)*

Exercise Saoirse Nua – the Irish-led NATO–Afghan joint-EOD operations.

Briefing UNIFIL HQ staff on logistical matters,
including food issues.

My dad and stepmother Betty in UNIFIL HQ,
Naqoura, Lebanon.

Briefing UNIFIL senior leadership on the capabilities
of the Irish-manufactured HOBO robot.

An aerial view of the explosion site at Flagstaff,
Border Crossing Number Two.

A briefing for Ban Ki-moon (then-United Nations Secretary-General) in the Defence Forces Training Centre at the Curragh on the Irish EOD capability and how it fits into UN operations. *(PA Images / Alamy Stock Photo)*

Conventional ammunition converted into IEDs. *(PA Images / Alamy Stock Photo)*

An all-wooden IED made to defeat the use of metal detectors (Afghanistan). *(PA Images / Alamy Stock Photo)*

Chemical, Biological, Radiological and Nuclear (CBRN) weapon training in Dublin Airport.

A CBRN training course using specialised kit.
(PA Images / Alamy Stock Photo)

The International Commanders' Course in Marauding Terrorist
Attack (MTA). An innovative concept developed and run by the
Ordnance School with the support of the Defence Forces, held in the
Ordnance School in the Curragh.

Maidan Square, Kyiv, June 2024. The flags sadly represent the significant loss of life in the war.

Captured/recovered Russian ordnance from the war in Ukraine (June 2024).

Innovative use of APOPO dogs in the identification of
mines and explosive devices in Ukraine (June 2024).

My beautiful grandchildren –
Amélie, Bobby, Ailbhe and baby Jamie.

With my family at my daughter Clare's wedding to Rory.

accepting invitations to dinner whenever they were offered. It was my job to get to know the locals, whether they liked me or not. I was also aware of the importance of the Catholic Church out there at the time and the considerable power it wielded. How it used this power was something that truly shocked me. I'll say more about this later, but as an example, I went to my local parish priest and I asked him if he'd mind mentioning to his congregation at all Masses that we were here to help, not fight. I went to Mass myself that Sunday and he never mentioned it. Little did I know, this was only the tip of the iceberg.

Our first opportunity to do real work came in the old Ottoman city of Mostar. Again, it's now a tourist favourite, and the old bridge is a renowned venue for diving the 70 feet into the Neretva River, but in 1992, it was under constant bombardment from Serbs in the surrounding hills, fighting the Croats for control of the city. I decided to do a risk assessment of the situation by driving with Volmer through Mostar as it was being shelled. My view was that the people didn't like us, but if they saw us driving through the town on a day when it was being shelled, they'd say, 'Well, maybe Europe is here to help.' We had to win these people over. It was down to us, really: the only other international organisation in the area at the time was the Red Cross. There was no UN, just us.

I had a beautiful, white, armour-plated Mercedes Land Rover with two big EU flags flapping on the bonnet. I had this wonderful driver from the Czech Republic, Tony Novotny, a marvellous guy, who could get us out of anything. So, we did our drive through the town, which was uneventful – until we were leaving, when there was a massive explosion behind us. Such was the pressure of the blast that we went from about 80 miles an hour to 150. How Tony kept the car on the road, I'll never know. I don't risk the lives of those who work with me, but I'm aware that I'll always push the boundaries of what's possible. It was a calculated risk, but with a purpose behind it. It worked, but the three of us could have been killed. Was it worth it? I'd like to think that it was.

Later that night, under curfew, I got a visit from a man I'd never met before. He said, 'Why do you spend all your time in the west of Mostar? You never, ever come to the east side.' He added, 'I saw you driving around during the shelling and I'm sure your Croatian people were very happy with that, but we weren't.' 'We' turned out to be the substantial Muslim population in the eastern part of Mostar. In our haste to get in on the ground, our briefings had been sketchy, so in my ignorance, I'd assumed the whole city was Croat with a small Muslim population. This man was here to tell me that half of the town, whom I'd never visited, were Muslim.

I thought, *right, we'll rectify that tomorrow.*

The next day, Tony and I walked onto the Stari Most, as the bridge was called. (When I was in Mostar, the bridge was still intact, but it would be destroyed two years later, blown to smithereens.) All was quiet as we crossed over the river, but just as we got to the other side, I received a sharp blow to the side of the head, and everything went black.

We woke up in darkness with very sore heads. It would seem that we'd been knocked out by rifle butts and dragged into a cellar. The door opened and an officer of the Muslim army came in, with a bottle of rakia, the local fire-water, in one hand and a gun in the other.

'I think it's over, Mr Lane,' Tony said.

'I don't think so,' I reassured him.

The light flicked on, and our friend was sitting at the table, like in a scene from a movie. 'Can you explain why, even though you've been in Mostar for such a length of time, you've never, ever bothered to come over here to see what it's like?' the man said.

I replied, 'I'm going to tell you the honest-to-God truth: I didn't know you were here.' The answer was so stupid, it had to be true. The man, whom I later came to know as Colonel Arif Pašalić, went out and came back with three glasses. He poured a shot of rakia into each and proposed a toast. We clinked glasses and drank. After more than a few shots, I promised him that I'd be back the next day and that, from then on, I'd spend 50 per cent of my time on the west side and 50 per cent on the east. And not only that, but I'd also ensure that there was Muslim representation in all the committees that I was involved in in the city of Mostar. This would later come back to bite me when the local paper published a cartoon of me embracing a Muslim person, standing in a pool of Croatian blood. That told me what some of them thought of me.

Tony and I wombled back over the bridge into Croat territory, where we were promptly arrested and brought to the police station. I knew that the Croats wouldn't touch me, because they wanted to be part of Europe, but they would certainly like to put the fear of God into me. So, I wasn't a bit surprised when the head of the military police came in, the aforementioned Valentin Ćorić. Now, military police in war are bad news – think of the SS in Nazi

Germany, or the military police in Argentina in the 1970s. Ćorić looked as if he fitted the bill, being over six foot and a former judo champion, but he also knew that he was wasting his time with me.

I said, 'There's no way that we can stand over what's happening in this city. I'm going to spend 50 per cent of my time over there and 50 per cent over here and if your people don't like it, that's tough. I can bring my ambassador here, but remember, this is Europe you're talking to.' The word 'Europe' gave me the ultimate leverage. Ćorić's boss, Jadranko Prlić, had said it to me one day: 'We want to be the brightest star on your shoulder, Mr Lane.' Croatia's eagerness to be part of the EU meant that I had some power in the situation. So, once I'd made my position clear, we changed the conversation and got onto chatting about our families, and I managed to swing an invite to dinner at his house the following Saturday night. In terms of intelligence, it was one of the best moves I've ever made.

I didn't want to be Ćorić's friend: I just wanted to disarm him and to find out as much as I could about how he operated and where the Croat forces were deployed and so on. In the informal setting of his home, Ćorić gave me so much information that Prlić stopped the meetings. That phrase about the banality of evil comes back to me when I remember Ćorić: a family man and a perfectly pleasant individual at his dinner table, yet who took part in the savage ethnic cleansing of the Muslim population. At the time, I found it hard to understand, but after a few months in Bosnia, I became familiar with the sad reality.

The next day, I went over to the 'other side' of the Stari Most with Colonel Pašalić, and I did a tour of his hospitals and schools, getting a sense of the living conditions for the Muslims, the whole picture. And it wasn't a pretty one, I can tell you. They were being starved by the Croats and shelled by the Serbs, both forces vying for control of the city. However, as a result of our visits, we succeeded in getting their electricity restored and food delivered to the Muslim side, as well as monitoring the Croat side of the city, just as we'd said we would.

One of my 'favourite' activities at this time was to go to the middle of the Stari Most and just stand there, as if I was taking in the scenery. I wasn't. I would simply have received word that the Croat forces wanted to blow up the bridge, which they wouldn't do with me on it. In my testimony to the later investigation into war crimes in the former Yugoslavia, I explained, 'To the Croat side, it was a bridge to a culture that they no longer wanted. To the Muslim side, it was a bridge to the western world which they needed to survive.'

This action, and our pledge to the Muslim population, came at a high cost to me and to the ECMM mission. Warnings were delivered on a regular basis. I went out to my car one day and noticed that the windscreen had been hit by one round of ammunition, exactly where my head would be if I was inside. The windscreen was made of armoured glass, but the message was clear: stay away from East Mostar. On another occasion, a plank with two nails in it was placed under the front tyre, so that it would puncture as soon as I drove off. When I took myself to the self-styled government, to Mr Prlić and his friends, asking them to do something, they feigned shock and surprise. I was determined to keep my people safe – but if the government had been hoping that their actions would affect the way we were going to do business in Mostar, they were wrong.

CHAPTER 10

QUEEN OF PEACE

I f you remember, I mentioned that the Catholic Church played a big role in the Bosnian conflict. During our time in Mostar, I had gone to see the local parish priest, Father Zovko. I asked him if at Sunday Mass, he would tell his parishioners that the ECMM was here to help them and should not be put in any danger. As I would write in my later recollections of events: 'There's no question that a word from him to the Bosnian public, to the Croats, would have assisted us, no question about it. Following that, our experiences in the whole area of Herzegovina, in the Bosnian Croat-controlled area, changed fundamentally. People ignored us completely. Once they heard that we had gone to the Muslims in East Mostar, to the other side, everything changed.' And the Church hadn't helped us in any way.

However, the true picture became apparent in Stolac, a tiny village close to the shrine of Medjugorje. We were being guided – selectively – by a Croatian colonel when I spotted a building on my left. Something caught my eye, something white flapping in the wind. It looked like there might be somebody living there. 'What's that?' I asked the colonel.

'It's empty,' he replied brusquely.

'Tony, drive in there,' I said, pointing to the building.

'No, no,' the colonel protested. 'There are mines there.'

'Tony, drive on there,' I insisted. Tony turned left and parked the car beside the building. It turned out that it was the remains of the village hospital, barely recognisable now. This building was in a dreadful state: the surrounding trees were half blown away from artillery shelling, the walls pockmarked with bullet holes, and behind the clotheslines with the white sheets on, blowing in the breeze, lay a row of freshly dug graves.

In life there are profound moments, when you stop and think, 'F*cking hell', for want of a better word. Unable to believe what I was seeing, I went inside, and what hit me was the smell of formaldehyde. The Croatian colonel shifted uneasily beside me. The next minute, from the shadows, people began to appear, everyone from a mother carrying a baby to an 80-year-old man. There must have been more than a hundred of them, all patients under the care of a doctor who introduced himself as Mehmet Kapic. He explained that their days would begin by hunting for any leftover food, because most of it was going to the Croatian army (90 per cent of the UN supplies of food to Stolac were going to the Croatian army, according to Robert Fisk). Then in the evenings they would hide from the Serbian shelling raining down on the village, before creeping out in the early hours to bury the dead. I was dumbstruck. To this day, I think about those people and the conditions in which they were forced to live.

My next thought was *we need to do something about this*. I told Tony to head into Medjugorje, to visit the town's parish priest, a Father Slavko Barbarić, who I'd met a number of times and who was close to those who had first seen the apparition of the Virgin Mary in 1981. For those of you who have visited the shrine, you'll know how busy it is, with busloads of people arriving and departing, gathering around the shrine to pray. However, because of the war, the town was completely empty. There were thousands of vacant beds and loads of food and water going to waste.

After a bit of small talk, I said, 'I've come for help'.

He smiled. 'Yes, Mr Lane, you're Irish and a Catholic like us, so whatever you need, we'll give you. Your country has given us so much. Where do you want us to bring it?'

'To Stolac,' I replied.

'But there are no people in Stolac.'

In my innocence or stupidity, I said, 'But there are,' and then I realised what he'd said to me. That they weren't 'people', they were Muslims. I was shocked but, sadly, not surprised. I had visited a couple of detention centres for the Muslim population, so I knew what was going on there, a situation which would later be described by the International Criminal Tribunal for the former Yugoslavia (ICTY):

In April and June 1993 two facilities were used by the HVO for the purpose of detaining Muslims from the villages around Kiseljak town, namely the barracks and municipal buildings in the town. The prisoners were initially detained in the barracks where they were kept in overcrowded and unhygienic conditions, their valuables having been taken from them. The prisoners were beaten regularly and kept short of food. Witness Y was transferred from the barracks to the municipal building which he described as being in a terrible condition, dirty, with a lot of garbage and mice running around: with 50 people to a room and no food for two days. The prisoners were taken to dig trenches on or near the front line. One dug trenches for a period of over eight months during which digging four prisoners were killed. Another was shot and seriously wounded while digging.

I decided that it was time to up the ante. As I was a diplomat, not a soldier, I was guided by the EU regulations, but part of my job was to brief the media on what was going on. Now, I could have simply passed press briefings on to my boss and left him to deal with it, but I interpreted 'briefing the media' liberally. Nobody had specifically said that I shouldn't talk to the press …

I heard that the famous journalist Robert Fisk was in Sarajevo. Now, when the ambassador had briefed me, he'd expressly told me not to go into Sarajevo, but I was determined to get the message to the journalists covering the war there that there were terrible things happening elsewhere in Bosnia-Herzegovina.

Getting into Sarajevo was a big deal back then, with an escort having to be arranged with UNPROFOR, but eventually, I found my way to the BBC studios, where the journalists were taking shelter from the Serbian snipers in the buildings around them and reporting on what they saw. There Fisk was, sitting on the floor of a makeshift office with the BBC's Martin Bell, playing cards.

'Well, how is Sarajevo?' I said.

Robert said, 'Ray, incoming, outgoing.'

'Look, if you want to find out where the real war is in this country, come down to Mostar,' I said, and marched out of the building. Message delivered.

Two days later, there was a knock on my office door and when I opened it, there was Robert Fisk. He'd thumbed a lift from Sarajevo to Mostar, which was an amazing feat, because of the many deadly checkpoints on the way. I was sure he'd had to talk his way down.

We sat down and chatted over a cup of tea, during which I made my position clear. 'I'm delighted you've come, but I have to be careful that I don't compromise my position here. So, I'm going to dress you in white and put the EU stars on your shoulder and I'm going to take you to four places. I'm not going to brief you on any of the four places. When we leave, I'm not going to ask you your opinion, and if you choose to write articles on those four places, that's your business.' He agreed.

The first place we went to was the hospital in Stolac. Fisk went in and met Dr Kapic. Now, just to paint a picture of how resilient these people were, the nurses at the hospital had painted their nails and put lipstick on and played music in the bunker to welcome him. It reminded me of all the times that Volmer had gone down to visit them, dancing away with them, the music echoing through the building. Now, nothing could disguise their situation, which was dire. A couple of hours later, Fisk emerged and we drove back through the checkpoints to the base in silence. Fisk said nothing until we sat down to have a drink with Volmer, and then Robert said, 'I've been in many places, Ray, but that was something else.' I didn't reply – there was no need.

Over the next few days, I took him to Kiseljak and Posušje concentration camps. They were called detention camps, but as you can see from the above testimony at the ICTY, they were, in effect, concentration camps. Fisk took it all in and wrote copious notes, before leaving a few days later. There was silence for the next couple of weeks, and then, on 17 December 1992, I got a call from Ireland. An article had appeared in the London *Independent* about a town in my area called Stolac. I professed ignorance.

However, it was Fisk's article on Medjugorje (also published in the *Independent*, 18 December 1992) that really got me in hot water. To his credit, he never mentioned my name: I was merely 'an inquisitive EU monitor', but I knew that he was talking about me. After all, I had brought him to Medjugorje, I'd introduced him to the visionaries who had seen the Virgin Mary as children and to Father Slavko. I didn't have to put the maths to him. He said, 'Ray, there are twelve thousand empty beds [in Medjugorje] and yet people are living in squalor. This doesn't make sense.'

All I could say was, 'Write it, Robert.'

When I talk about Medjugorje, my problem is not with the many pilgrims who flock there every year. Rather it was the use of the shrine as a bastion of extremism during the Bosnian War. The apparitions in 1981 and the religious fervour that came with them were exploited by certain elements, whipped up so that people would be prepared to do anything to defend the faith. One of the Croatian army leaders was a man called Milivoj Petković, whose Kalashnikov was emblazoned with the cross of Medjugorje, a set of rosary beads around his neck. He would later be sentenced at the ICTY for the crimes of ethnic cleansing and keeping prisoners in inhumane conditions. For Father Slavko, the shrine of Medjugorje was more than just a place to gather; it was a symbol of extremism, one that we've seen all over the world.

In the aftermath of the article, there were many phone calls to the press office, conveying the Irish army's disquiet at Fisk's depiction of Medjugorje, at his question, 'Why has the Christian church here not opened its heart to its Muslim brothers?' One person wrote into the London *Independent* querying Fisk's assessment of the situation, saying that Medjugorje was acting as a hub for the distribution of international aid. All I can say to this is that I know what I saw and what I heard there, and so did Robert Fisk.

The day before I would leave to return to Ireland when my tour was over, I decided to go back to talk to Father Slavko. He was a really forceful man, with a powerful personality and sharp eyes which seemed to bore into people. I made sure to rehearse what I wanted to say to him. I knew that if we were just sitting face to face, I wouldn't be able to talk to him, because of the sheer intimidation of his presence. I'd seen him in action on many occasions, and indeed, one of our disagreements had left his secretary, Marina, in tears.

I can still remember her turning up at my office one night after curfew. I was astonished. 'Marina, you could get killed out there. What are you doing?'

She'd stood there nervously, before blurting. 'You're wrong about Father Slavko. The man is a saint.'

I didn't share her view, instead ushering her into a chair. 'Where do you come from, Marina?'

'Montenegro. But all my family are dead,' she said.

I was horrified. 'They're dead? How did that happen?'

She shook her head. 'I mean, in the eyes of the Church they're dead.'

It was utterly chilling to witness the brainwashing of this woman to reject her Muslim family in action, so when I arrived at Father Slavko's residence with my translator, Alenka, I knew what to expect. Of course, he didn't want Alenka to come in because she was Muslim, but I insisted that she stay. My message was simple: I went through the Ten Commandments and how I felt he was manipulating them. 'Thou Shalt Not Kill, unless it's a Muslim.' 'Love thy Neighbour ... unless they're Muslim ...' and away I went, going through each commandment. I was aware that his hands, which he'd been clasping together, were getting whiter and whiter with every line I spoke.

When I was finished, he said, 'Mr Lane, we didn't really get on well when you were here, but you don't realise that I'm the shepherd of a large flock of white sheep, and around those white sheep are black ones, and it's my job to keep those white sheep safe.'

I went out to my car and climbed in. I rolled down the window to let in a welcome breath of fresh air. Father Slavko reached in and extended a hand to me. I didn't respond. It was the only time in my life I'd ever refused to shake anyone's hand. We drove away and I never saw the man again. Until recently, I was unaware of the fact that Father Slavko was illegally staying in Medjugorje, having been asked to leave by the bishop on more than one occasion.

You might wonder what my actions achieved for the people of Stolac. Well, once their diverting of UN food was revealed, the Croats sent through more food to the besieged Muslims, and the Serbs reduced their shelling of the location, that's for sure. The people in both detention centres started getting their proper rations. Perhaps the pen really is mightier than the sword.

The next time I met Robert Fisk I was in Beirut Airport during my 1997 posting to Lebanon, getting ready to catch a helicopter down to my HQ. I was carrying a tray of eggs – I'd had a tender for eggs and these were samples. As I was heading for the helicopter, Fisk was disembarking. He walked by, looked at the tray of eggs and said, 'Ah, right, this is what you've become.' We laughed heartily at that, and that was the last time I met him.

Fisk was an amazing journalist. As capable of egotism as any headline-grabbing foreign correspondent, he had the knack of digging around and finding the stories that others just weren't able to grasp. During my time in Bosnia, I hosted many journalists as part of my job, everyone from Dan Rather to the BBC's Kate Adie. I tried to accommodate them as best I could, even if sometimes their recklessness (as far as I was concerned) rubbed me up the wrong way. If I thought they were putting my monitors or the locals in danger, I would put them straight.

One foreign correspondent, who I won't name but who stayed in my house with my team and who I spent some time with, was particularly problematic. I was driving this person around the area and I pointed out a town to them, saying, 'You can't go in there. Don't even write about it.' The stakes were simply too high. Of course, it was like a red rag to a bull and the journalist charged in there, eager to get the story. Two days later, Valentin Ćorić rang me. 'We have your media person in jail here, Ray. We caught them in the town.' They'd roughed them all up a bit, but any pity I might have felt for them was outweighed by anger. 'I told you not to go there and you did. Let me tell you what that means in my little world: I'm going to get a phone call over the next six months when they've done something horrendous and they're going to say to me, "Mr Lane, we looked after your TV crew…". I hoped that the message had been received. Sadly, what I'd predicted happened. Three months later, I got a phone call from Ćorić to tell me about some horrific incident: 'And we'd appreciate if you didn't report it,' he wheedled. My hands were tied.

However, one Irish journalist sticks in my mind because of her sheer bravery. Valerie Hanley was just a rookie, but she paid for herself to get all the way to Bosnia, complete with packs of sausages for the team! I took her to Gornji Vakuf to report on it, and she got an exclusive. The issue was that she had no way of phoning in her story, so I said to Kate Adie, who was also in the town at the time, 'Can you lend her your sat phone?' Being the BBC, they had all the equipment. She merely looked at me imperiously and said, 'My dear, you seem to miss the point here. I'm the person who's reporting on this, not anybody else.'

*F*ck you,* I thought. And, even though it cost a fortune, I went out to get my own sat phone and drove around until I could get a strong enough signal for Valerie to phone in her story. She did. She got her exclusive and I was delighted to have helped her. Yes, I should have been objective, but being objective in this situation had to be outweighed by my humanity.

CHAPTER 11

'NOT WAVING, BUT DROWNING'

Gornji Vakuf, 1992

'Do you know what's happening in my town, Gornji Vakuf?' I hadn't even heard of Gornji Vakuf until one of its citizens came to me and asked me that question. I didn't know, but I decided to find out. Along with Tony, my ever-reliable driver, I headed to the town very early one morning. We perched on a hill overlooking the town, which was being pounded by artillery fire. It seemed to me that the firing was coming from Croat positions. I decided to go back to Mostar to talk to the key leaders down there, so they would talk to their key leaders in Gornji Vakuf and put manners on them. It was time to pay a visit to Jadranko Prlić, leader of the self-declared Croatian Defence Council. He was a suave and intelligent guy, a professor of economics, who would later receive a 25-year sentence at the ICTY for crimes against the Bosniaks.

He said, 'You're bringing nothing to me, Mr Lane. Not even plans. I mean, I don't even know if my people are up there.'

'I can guarantee that they're up there,' I replied. With Prlić unwilling to help, it was time to bang some heads together. I decided that I would go back to the town and see if I could get the sides to come together to agree on a plan to restore peace to the town. I went back to the office and there was Alenka, case packed, ready to come to Gornji Vakuf. I'd promised her father that I wouldn't put her in danger, ever, so I said, 'No, Alenka, I can't take you.'

'Well, who's going to be your interpreter?'

'We'll get somebody up there.'

She said, 'No, I'm coming,' and before I knew it, she was in the back of the car with her suitcase. She was some girl. The only Muslim interpreter of my 11 in total, we'd found her by accident, hitching a lift on the road. The great thing about her was that with her name, the Croats weren't sure if she was Croat or Muslim, so she was a very valuable part of the team. She used to sit beside me at meetings and when I'd start waning and get sick of it all, she'd start kicking my leg. 'Mr Lane, we're nearly there. Do you realise how close we are? Keep going.' If anyone gave any cheek in our meetings, she would machine-gun a reprimand in Serbo-Croat. I'd often say, 'What was all that about?'

'Never mind,' she'd reply, with a smile. Without Alenka, I know that so much less would have been achieved.

When we got to Gornji Vakuf, we hired a meeting room for the various sides, first making sure that all the weapons were kept outside. To begin the meeting I said, 'I'm going to ask the parish priest to say a few words, then the imam, to guide us.' The parish priest got up and he had a Bible in one hand and a Koran in the other. He put the two together and I thought, *we're starting off well.* Then he began, 'But in 1648, when the hordes of Muslims came through here, they showed no mercy to our people.' *For God's sake*, I thought, *I specifically asked him not to make any political statements.* I looked at the imam and I said, 'If you're going to do the same thing, just skip it.' But he made a wonderful speech, about their shared past of working in the fields together as neighbours and friends.

Over the next 14 consecutive days, for 10 hours a day, we hammered out an agenda for a potential ceasefire. The items that needed to be resolved included burial of the dead; burial of animals; restoration of essential services, such as electricity, schools and so on; and equal distribution of humanitarian aid. After three days, it was going nowhere, and I was getting frustrated with the impasse. One evening, in my office in Mostar, I got a big sheet of paper and wrote in English 'RULES OF THE MEETING'.

Rule number 1: No use of the word 'but'.
Rule number 2: No history lessons.
Rule number 3: No hidden agendas.

I got it translated into Serbo-Croat and made three large copies. And when they came in the next morning, I said, 'Gentlemen [there were no women, apart from Alenka], these are the rules, and if somebody breaks the rules, I'm just going to tell them to sit down.' Who was I to tell them what to do in their village, you might ask, but when it came to getting agreement, I had to be tough if I was going to achieve anything at all. In fairness, they cottoned on quickly and if anyone broke the rules, it would be pointed out to them.

Getting the meetings to operate successfully took every ounce of my military training. When I was a lowly cadet, we'd used to talk about the concept of 'negative process', that is, meetings that achieve no outcome but are simply there to be meetings. We've all had a few of them! As an aside, I read an interesting book not so long ago called *Moving Mountains* by General Gus Pagonis, and I picked up a few tricks, which I use to this day. Pagonis was director of American logistics during the Gulf War of 1991, but he never held a meeting across a table. All Pagonis's meetings were held standing up. I was familiar with the concept that the British Army referred to as 'car-bonneting', i.e. holding meetings across the bonnet of a car. I'd found it very useful at times in Bosnia. It's genius, because nobody has the time then to get sidetracked.

Another trick of his was to distribute a set of index cards to those present. If there were any comments or questions, they could be written on the card and thrown on the floor. He would pick one up at random and answer the question on the card. Once, however, another card was added in: 'Some people say that General Pagonis doesn't listen.' He was holding the card and said, 'Okay, that's a bit of a surprise. Would anyone like to add something?' Whereupon one of the attendees said, 'Now, sir, I didn't write that card but there are times that you don't listen.' Who wrote the card? Pagonis.

Process is important and I'm a great believer in it, but it has to be positive. In Gornji Vakuf, the emphasis was on 'positive process'. We were meeting when local people were in the middle of horrendous situations, but the meetings we had were structured so that all the peripheral stuff, like what happened in 1322, was pushed away: instead, we were just going to concentrate on the agenda we had in front of us, and it worked. We just got them to concentrate for a few short hours on what was written in front of them, and before long, they were taking ownership of it. After a while, if anyone new came to the table and started on about history, they'd be told, 'No, no, look at rule number two …' Our goal was to live completely in the moment and to aim for concrete, achievable goals: for example, we'd say, 'If we can get the food sorted, or the electricity, or the water then we'll deal with the bigger picture.' Of course, it doesn't go away for ever, but that's not necessary. What is necessary is a focus on what can be done right now. Of course, there are always exceptions to the rule, as I learned later on in the process.

We went through all the items on the agenda until we came to the last, which should have been the easiest: equal distribution of humanitarian aid between the Croats and the Muslims. I wanted there to be one centre from where the aid would be distributed equally. There would be no preferential treatment, I thought.

This man came into the meeting, and he stood up and said, 'You cannot ask my people to share food with these people. You fail to understand what's happened here. You fail to understand history.' I pointed to the board and its mention of 'No history lessons', but he didn't even look at it. 'It won't happen, Mr Lane, do you understand that?' Now, he was an older man, so I couldn't simply dismiss him. A Muslim, he had status in Gornji Vakuf, because he'd run a large souvenir business in the town during the good old days. So, when we had a break for coffee, I went down to talk to him.

'I don't envy you your job,' he said. 'In fact, I have no idea how you do it. But I cannot and will not support this.' I had no idea how we might move forward on this basis, but then he said, 'Would you come out and visit me tonight?'

'Out to your house?'

He looked at me carefully then said, 'I'll give you directions to where I am.' That night Tony and I drove out to his location, but there was no sign of any house. All that we could see in the beams of our torches was mud and earth. The next thing, a head popped up out of the ground, like a mole. He was living in a homemade bunker in the garden of what had once been his house. He led us down a set of steps into his underground burrow, where he offered us rakia and took out the family albums to show us. The town of Gornji Vakuf looked idyllic in the photos, and he explained that Muslims and Croats used to go to weddings and Christmas festivities together, in a town that was pretty equally divided between Croats and Bosniak Muslims.

'By the way, where are your family?' I asked him after a while. I'd only noticed his own clothing and essentials in the bunker.

'They're dead, Mr Lane. The people you want me to share food with, they killed them.' I had nothing to say. I was trying to get the two sides together, but how could I get past this reality, I wondered? I wanted people to share, but it struck me that I knew so little about what they were enduring. As I was leaving, he said, 'I want to give you a present.' He gave me a calendar made of wood and said, 'That was made in the factory here and I'd like you to have it.' I thanked him and went back to the office and didn't sleep a wink.

The next morning, I went to the hall to finalise the last item on the agenda. The locals were all there waiting. 'It's a very difficult decision to make and I understand everybody's perspective here. I've tried to be sensitive to people's viewpoints; however, the decision is that we will store food together and we will equally distribute it.'

I looked at the man to gauge his reaction to my decision, and he was in tears. It was a bitter blow to him, I could see that, so I went over to him to explain. 'Listen, I had to make that decision.

I'm actually not happy with it myself, but truly, to keep people safe here, to keep some sort of peace, this is the only way out.'

He said, 'I disagree with you fundamentally, but I respect you.' And that was it.

We managed to implement the decisions of the committee in Gornji Vakuf, with some exceptions. They couldn't agree on the burial of the dead, so we picked a soccer pitch, which we divided in two. One side was the Muslim side and the other was the Croat side, although there weren't too many in that. We organised an interdenominational service for burying the dead. I had to pay a reluctant visit to the parish priest again to get him to take part, but I made it clear that I wanted him to behave himself. At one end of the pitch lay the remains of the changing rooms, and as I was interested in soccer, I went over to take a look. I could see that the team pictures had been ripped off the wall and smashed on the ground, but I picked one up and examined it. In it, the commanders of the Muslim and the Croat sides were standing side by side in the back row of the photo. They'd been on the same team and had played together for eight years.

I couldn't fathom the depths of the hatred between the two sides in Bosnia. Burying a dead child in frozen ground on 3 January, as we did in 1992 – you're not normally meant to do that. After one discovery in a local barn, which is too disturbing to share here, I asked Valentin Ćorić a question. 'Tell me, what I saw in that barn, my God, I can't believe that Europeans would do that.' That was when he gave me the answer that opens this section of the book. 'Out here, Mr Lane, we do that [mimes pushing soil away with his hand] and we're barbarians. In your country, you have to dig a little bit deeper.' He was right. All the bitterness and resentment were just under the surface, bubbling away, ready to ignite at any moment. They still are.

With all the items on the agenda now sorted, it was time to organise the peace agreement for Gornji Vakuf. We drew it up, with timings for the withdrawal of the Muslim and Croat forces, along with their mortar positions and tanks. Then we got together the head of the local UN peacekeeping force; the leader of the Croatian army, General Željko Šiljeg; and Arif Pašalić, the Muslim commander. All they needed to do was to sign on the dotted line. Instead, Šiljeg, a big bull of a man, threw it at me. 'Mr Lane, you give the impression to everybody here that you're a diplomat. I know that you're a military man. And no military man on the verge of a major victory is going to walk away and retreat.' Then he added the body blow: 'There'll be no signature, there's no treaty.'

'Fine,' I said. It was time to take the gloves off. 'The next person who dies in Gornji Vakuf, I will hold you personally responsible. I'll get on to the ambassador in Zagreb, who'll go on all the television channels around the world and say that the EU were on the verge of a unique peace deal here and you stymied it.' Strong words, but I wanted him to get the message. He stormed out. The peace deal looked as if it was dead in the water.

The next day, a child was out playing, and he stepped on a mine, dying instantly. By this stage, Šiljeg was in Sarajevo in a meeting with the head of UNPROFOR.

'Tony,' I said, 'We're going to Sarajevo.'

When we got there, I invited myself into the meeting and placed my hand on Šiljeg's shoulder. I said in a low voice, 'Remember I told you about what I'd do after the next person who dies in Gornji Vakuf? The ambassador in Zagreb is going on the media in one hour and is naming you as the murderer of Gornji Vakuf.' And I walked out.

My deputy said to me, 'I think my career just ended there.'

'No, I think you'll find it was mine,' I replied.

We were on the road back to Gornji Vakuf when a swarm of armoured vehicles passed us on the road. When we got back, there Šiljeg was, maps of the area out on the table, ordering both Muslims and Croats to move their positions. Then he looked at me and he said, 'Now, Mr Lane, are you happy?'

I said, 'Why didn't you do it yesterday?'

Before the peace treaty was signed in Gornji Vakuf, we arranged for an exchange of the bodies claimed by both sides. We picked neutral territory – a road outside Stolac – and I drew white lines on the road with a 15-foot gap between them. I'd seen this process carried out by the International Red Cross and, in my naïveté, thought it would work here. The Croats would stand behind one line and the Serbs behind the other. We had a list of names from both sides, and we had a table laid out with food and we'd asked an Orthodox and a Catholic priest to attend.

The two trucks containing the bodies pulled up and I said, 'Okay, first body out.' So, the Serbs opened up the back of the truck and a head rolled out of a black plastic bag. A Serb kicked it back into the bag. The Croats saw this, of course, and they took a body out of their plastic bag and burned its eye sockets with a cigarette.

I lost my temper. I told them to put their bodies back inside and that it was an outrage against God, Christ and Allah. 'This is inhumane. You're not civilised people. Put your bodies back in and go back to your ghettoes.' Hardly well-chosen words, but I was livid. All the trucks rolled out, but two weeks later, I was asked if I'd do it again and this time, it was much more civilised. We bagged the bodies and the priests said prayers and those who had died were returned to their families.

Who is it who said, 'history poisons everything'? We in Ireland should know that better than anyone. But I simply hadn't been prepared for the savagery in Bosnia. I handled what I was seeing with difficulty. I was working 20 hours a day, seven days a week. I was physically and mentally exhausted, scarred by what I was seeing every day. So was everyone on my team and we all coped in different ways. Music became something that helped us, and we developed a dependence on a particular cassette tape that we couldn't do without. The choice of artist didn't matter, I was a fan of Clifford T. Ward and wouldn't drive without his song 'Not Waving, but Drowning' playing loudly in my ears. My driver, Tony, favoured an Italian singer called Eros Ramazotti, and another one of the team played Annie Lennox's 'No More I Love You' on repeat. The key was that this random selection of songs seemed to sum up the situation that bound us together. The cassette would accompany every trip we made in the car, and so precious did it become that we'd lock it away after each trip in case it disappeared. The mind does funny things in situations like these.

In the middle of my posting, I returned to Ireland to be best man at my father's wedding to my stepmother, Betty. Mammy had sadly died of cancer in 1986 at just 57 and Daddy had been completely lost, so when Betty had appeared in his life, we were all delighted. She was a wonderful woman and brilliant for him. I was very fond of her and had been honoured to be asked to be best man. Daddy picked me up at the airport and we went to their new apartment in Dun Laoghaire. Betty was showing me around, pointing out the view and the features. I suddenly felt my stomach clench and had to run into the bathroom, where I got sick. I couldn't take the normality of it all, after what I'd left behind in Bosnia. It just seemed bizarre that everyone in Ireland could be going about their normal business while there was a war going on.

I'm ashamed to say that I went through my duties at Daddy's wedding like a robot, although I managed to make a speech, with great effort. Then I sat down and didn't talk to anybody for the rest of the evening. At the end of the reception, Bridget said, 'Ray, I think you should go back early [to Bosnia] and come home in May and have this out of your system because this isn't good.' I knew what she meant: I needed to leave Bosnia where it was and return home to them. I also knew that my family was paying the price for my mission to Bosnia. Every day when she went to school, my daughter, Clare, would hear me on the radio, telling people about the war and about what I was seeing. She used to call me the Gay Byrne of Bosnia. I was aware of the irony that her contact with me at this time was through the radio.

PTSD wasn't really a thing in my time, and our understanding of the condition has changed a lot in recent times. Back in 1992–3, I had nobody to talk to. I had to process these horrors by myself, as we all did on that mission. At the time, I felt that the army was more concerned about what I was saying about Medjugorje than what was going on in Bosnia. In the end, after my tour, I came home for one month's medical leave before beginning work again.

Two things saved me: running and a treasury tag.

When I returned to Ireland, I was exhausted, but my mind was going at a million miles an hour. While Bridget and the children got on with their daily lives, I was a ghost in the house. I'd get up in the morning and I'd run and run until I couldn't run any more. Then I'd get back and go to bed; then in the afternoon, I'd get up and go running again, hoping to chase the memories away. I was looking for exhaustion, so that my mind would stop tormenting me. I have no idea how Bridget put up with me – she should have chucked me out of the house.

After a month, I returned to the green uniform and to the ammunition depot in the Curragh. There was no debrief. There were no questions about how I'd got on and what I'd been through. Then one day I was in my office and the phone rang. It was my boss, asking to see me. I drove over to his office building, marched in and saluted him.

'Captain Lane,' he said, 'before you left Ireland, you submitted a report to me. The report itself was grand, but I was looking at where you'd punched the holes for the treasury tags and it was all wrong, so I want to show you how to do it properly.' He put my report on the table, then he took a ruler, and he measured one inch in from the edge of the page, slipping the tag through the hole. The process was repeated with the hole lower down the page. 'Now, he said, 'do you understand that?'

'I do,' I said.

'That's grand,' he said.

Believe it or not, that was a turning point. It was so silly and so banal, it made me realise that I was lucky to be home, to be in a place where people could fuss about the placement of treasury tags. I got into my car and started laughing, then I started crying, then laughing again. As I drove back to my office, I actually started feeling a bit better.

CHAPTER 12

ASYMMETRIC WARFARE

Afghanistan, 2007

'To stay alive in Afghanistan, it's 70 per cent tactics, techniques and procedures, 20 per cent technology and 10 per cent Murphy's law.' I explained this to my boss, General McNeill, in 2007, during an early-morning briefing. You can't plan for everything, for sure, but you can certainly plan for 90 per cent of what could happen, and the 10 per cent you haven't planned for is covered by training and by situational awareness – i.e. that you're so tuned into the prospect of the abnormal that you will be able to deal with it. This has always been my philosophy: that well-trained soldiers can deal with more or less anything. In fact, I believe that well-trained people can spot things that are out of the ordinary with the right training. You and I can be very effective in spotting things that don't feel right and, importantly, acting on that instinct.

The first time I went to the country was in 2002, 10 years after my posting to Bosnia and five years after Lebanon. I'd learned a huge amount from both of those trips, good and bad. In 2002, the Taliban had been conquered by a combination of the US and anti-Taliban Afghans. The US had begun a huge reconstruction project and had appointed Hamid Karzai as interim leader of the country. The atmosphere was almost festive. The Taliban had just left, and the locals were jubilant. There were only three of us in the team, and the purpose of the visit was to see if Ireland could offer our IEDD capability to the forces that were out there.

NATO hadn't taken over yet, so the force was run by the British. Thus, from the Middle East, we boarded an RAF C-17 plane bound for Kabul. The plane was enormous, a big beast of a thing filled with 150 men, various huge pieces of equipment and enough rations to feed an army for six months. I couldn't believe that something of that size could get into the air! It did, and as soon as we were up, the seasoned British soldiers unbuckled their seatbelts and pulled their sleeping bags out, and proceeded to lay them wherever there was room, including under the armoured cars. They obviously knew the importance of getting some sleep whenever they could. I learned a big lesson then.

As we approached Kabul, we were flying at about 30,000 feet, when the pilot's voice came on the intercom: 'Code red in five'. My stomach dropped to my boots. I looked at the nervous British soldier sitting beside me, and I said to him, 'Do you know what code red is?'

'No.'

'I haven't a f*cking clue either.'

We found out very quickly. The lights in the cabin went out and from 30,000 feet, this enormous plane dropped like a stone to about 10,000 feet, to avoid enemy fire. Then we hit the runway with such force that we bounced back up again like a tennis ball, until finally we taxied into safety and my stomach rejoined the rest of me.

After that, my first impressions of Afghanistan were good. We were driven into Kabul to our base in ordinary vehicles, easing our way through the busy city traffic. The people looked at us with smiles on their faces, waving in welcome, clearly glad to see the back of the Taliban. Our base in Kabul was a tented village and we were shown to our tent with our three beds in it and given shower bags, which you'd hang from the bathroom ceiling, then pull a cord and it would empty over you. It was all very civilised, if a bit basic. The only precautions we had to take were to make sure that our boots were turned upside down at night so that the scorpions wouldn't crawl into them.

The British Army couldn't do enough for us, and we soon settled into the rhythm of our visit, heading out to different locations to talk to people about the IED situation, then coming back at night and wandering into Chicken Street in the city for our tea. The locals knew what the army boys liked, so chicken and chips were on the menu, and even a beer or two. One of my colleagues decided to test out one of the local bikes, borrowing one from a local man, climbing on and speeding down Chicken Street. The Afghan man was waving and yelling something we couldn't understand until he finally managed, 'No brakes!' My colleague dug his lovely Irish boots hard into the rubble and missed a wall by inches. We all had a great laugh at that.

We had to remind ourselves that we were there for a purpose, so one morning, we headed to a British Army base outside Kabul to talk to the CO about the number of IEDs in his area. However, all he wanted was to talk about his hobby, which was photography. He'd taken magnificent photos of the Afghan landscape, which he was keen to show me. Eventually I had to say to him, 'Colonel, can we just sit down and talk about these IEDs?'

'Ah, no, Commandant, there are no IEDs out here,' he said confidently.

I let that slide. 'Have you any capabilities for dealing with them?'

'No,' he said cheerfully.

As if on cue, I spotted a truck outside the window. A ramp was descending from it and what should appear only a British Army bomb-disposal robot. I said, 'There's a bomb-disposal robot there!'

'Jesus,' he said, 'I didn't know we had one of those around here, because there are no IEDs.' It would seem that our trip was somewhat wasted, but it was still very educational. It was great to speak to the locals, who were mad keen to talk about Ireland. When I left, I remember thinking, *God, maybe there is a future for this country.* After that, I simply came home, wrote a report about it, and that was the end of it. I thought I'd never get back to Afghanistan, so my one and only visit had been worthwhile.

———◆———

Of course, I was wrong. In 2007, I was 54 and reaching what I thought was the end of my military career when I was asked to go to Afghanistan again. The British were pulling out of Kabul to go to the southern province of Kandahar, and they needed a COO to run the counter-IED branch. The British Army had recommended that the Irish Defence Forces' EOD people take over this important role. What a tribute to our professionalism and experience.

In the few years since I'd been there, the situation had changed dramatically. In spite of a general election and with a US-backed president, Hamid Karzai, in charge, the Taliban were still waging war in the south of the country. And their tool of choice was the IED. Furthermore, they were prepared to die in suicide attacks against NATO troops, who were operating under the umbrella of the International Security Assistance Force (ISAF), an international force whose job initially had been to protect Kabul, but whose area had increased to take in the whole country.

The problem seemed to me to be that the Americans were still thinking about war in symmetrical terms. Let me just set a bit of context here. Remember the Boston Marathon attack in 2013? That was an improvised explosive device – a pressure cooker packed with explosives. Or the Manchester Arena bomb, set off by a lone suicide bomber, to utter devastation. All that's needed in a climate of suspicion such as Afghanistan is for one person who doesn't like the peacekeepers to start manufacturing crude IEDs. That's what we call 'asymmetric warfare'. 'Symmetric warfare' is the rows of tanks with soldiers pushing forward against each other, just like they did in the Second World War.

In Afghanistan in 2008, 55 per cent of NATO casualties were from IEDs, and the Taliban had manufactured in the region of 50,000 devices. However, I used to fly down to Kandahar and see great patches of dry earth full of tanks, all parked up, doing nothing. The Taliban hadn't got any tanks and yet there they were, killing and injuring NATO soldiers and civilians. In US Army briefings, they relied heavily on drones. Each drone had a bird's name and at times, it resembled an aviary. 'We put the raptor up last night,' someone would say, or that a B-1B bomber had dropped 10,000kg of ordnance somewhere the previous night. The Taliban had no bombers either, but what they did have was IEDs and an innate knowledge of the terrain. We were fighting a battle that we had no chance of winning.

So, back to Ireland. I was stationed in Dundalk and was in a pub in the town on a Thursday night having a pint when a colleague happened to mention the position in Afghanistan. (Even though the Troubles were long over, I still made it my business to keep abreast of any developments in the area, as the terrorist threat was still live.) I was specifically asked to go, and yet I couldn't help thinking, *at my age?* I'd been to Bosnia, Lebanon and Afghanistan already and I'd seen just about everything; frankly, I wasn't sure I had another trip in me. So, I rang Bridget for some advice. I knew that I could always rely on her to put it to me straight.

She listened and then said, 'What are you getting out of this, Ray?'

She had a point. I had nothing left to prove, after all. So, I thought for a bit, then rang my boss back and said, 'I'll go on one condition.'

'Right …' he said cautiously.

'I'll go to Afghanistan, provided that when I get back, I be considered for a posting to Belgrade.' I knew that there was a job going there with the Organization for Security Cooperation in Europe, the OSCE, which might not be too taxing, but that would allow Bridget and me to travel a bit. Little did I know, I'd spend a further 11 years in the forces, without a sniff of a posting in Belgrade! That's a soldier's lot, and I have no issues with it.

'We can't promise anything, but we'll do what we can,' was the answer. *Fair enough*, I thought, accepting the honour of being the first Irish officer to go out to Afghanistan in this role.

This time, Kabul was completely different. There was no local driver, no messing around on bikes on Chicken Street. Instead, we were greeted by NATO soldiers and rushed out of the plane to armoured vehicles, to be driven at speed to headquarters. As we drove along, I could feel that the atmosphere was very different to

2002. The locals were now looking at us with hatred and the tented village had been replaced by a high-security military complex.

I had hardly put my bags down when a US sergeant major appeared at the door. 'Sir, just a reminder you're doing the CUA in the morning.'

I had no idea what the CUA was, but I didn't want to appear ignorant, because the way he said it, it was as if I was going to St Peter's in Rome to meet Pope Benedict himself. Eventually, I said, 'The CUA?'

'The Commander's Update Assessment.'

'Oh yes, of course,' I said. I had some awareness of NATO's global briefing system, which was all lights, camera, action and came with a rehearsal to boot.

He hovered in the doorway. 'You'll need to go to the command centre now.'

'Ah yes,' I said, trying to sound as if I knew what he was talking about. I knew that there were 125,000 NATO troops in Afghanistan by this stage, but I wasn't prepared for the scale of the centre and the sheer size of the purpose-built screening room – a huge auditorium with a vast screen, on which were a variety of pictures of the country being beamed live. Hundreds of people were working at computers in front of it. I could see members of the Air Force, Navy, Infantry, Artillery, Armour, Special Forces: they were all there. This was a real war.

I went to introduce myself to the counter-IED officer, who was Italian, and I said, 'I'm doing the CUA brief in the morning.' He was shocked because I was just in the door and I was going to brief my ultimate boss, General McNeill, a four-star general, on the IED activity in the past 24 hours. (The position of commander ISAF rotated between the member countries, with rotations lasting between six and eighteen months.) Nonetheless, my colleague helped me with my PowerPoint presentation, and I thought I was all set. I was suffering from that classic beginner's ignorance of what lay ahead of me.

The next morning, I was up bright and early at 5 a.m. and I made my way to the command centre again. The German Chief of Staff took the rehearsal, and we lined up in front of the audience, all of us: from the Air Force, the Navy, Special Forces, media and counter-IED and so on. When I spotted the US ambassador, the penny dropped about what a big deal this was. And it wasn't even the real thing yet. I was going to brief the audience on a country I knew little about, having only been there once five years earlier. When the German pointed at me, I stood up. He didn't even say hello, just listening to one or two of my points, then saying, 'Okay, that'll do, next please.' A female officer stood up to give the PR briefing and all he said was, 'Keep your comments short and to the point, and do not move off this X,' pointing to a large 'X' marked on the floor. The briefing was being filmed and the X ensured that we wouldn't wander off screen. The rehearsal done, we all took our seats facing the rows of chairs full of NATO personnel.

The next moment, there was a hushed silence and General McNeill walked in. The Commander of ISAF was of small stature but wore a very big hat. Trim and neat, his bright blue eyes missed nothing. He told us all to sit down and the briefing began. It didn't start well when the poor PR expert moved off the X, making her point by moving out of the frame, but more worryingly, moving closer and closer to the general, who was getting visibly agitated. The German intervened briskly. 'Moving on, counter-IED, Colonel Lane.'

I made sure to stand on the X to deliver my assessment of the previous 24 hours. I had stayed up late the night before to memorise the presentation so I wouldn't make any mistakes, and off I went through my itemised list of IED attacks, all marked on a map of the country. 'Item 1: IED attack Kandahar,' and I described the event in brief, then onto 'Item 2, item 3, item 4 …' I arrived at item 11, then 12, and was rolling onto 13 when McNeill stopped me. 'Colonel Lane, where's item 13 on the map?' I looked up: there was no item 13 on the map, and there was no item 14 on the map either. I swallowed nervously and said to the general, 'Sorry about that, sir, I'm not going to make excuses.'

He looked at me intently. 'Okay, Colonel, what part of the world have you come from, Italy?'

'No, sir, I'm from the Republic of Ireland.' He looked slightly taken aback, I'm assuming because he was of Irish descent. Then he said, 'Okay, people will tell you, Colonel, that I don't give two chances to people out here, so you'd better get it right by tomorrow morning.' That was it.

I sat down, shaking. I happened to be sitting beside a French colonel and a British colonel and the French guy whispered to me, 'You're going home.'

I was devastated. It might seem like a small thing, but every detail counts in this kind of briefing and the fact that I'd got off on the wrong foot occupied my mind for the rest of the day. It didn't matter who had left items 13 and 14 off the map, because it was my job to take the rap, but all I could think of was having to go through another CUA the following morning. This time, I made sure that every single item was on the blessed map! I gave a confident brief and didn't omit a single detail. This time, McNeill looked at me and he smiled, and said to me, 'What part of Ireland do you come from?' and I told him that I was a Dubliner.

'Do you speak Irish?'

'I do, sir, a bit, yeah,' I fibbed. Off he went with his own *cúpla focal*. I had no idea what he was saying, but I was delighted that I'd got away unscathed. I sat down and listened to the other briefings, and when McNeill got up to leave, we all followed him. He turned at the door and looked at me and said, 'I want to see you for a few minutes.'

I was dreading getting another dressing down, but instead he said, 'Apologies for not understanding your tricolour yesterday; that was appalling, nearly as bad as your briefing.' And he laughed. 'You've no idea how welcome you are here,' he continued. 'I rang my wife in America, and I said, "I was briefed by an Irish Army Officer today, how good was that?"' The power of the Irish still held in the United States, I thought, thanking God for the tricolour on my arm.

One small problem that we did have with our uniforms, however, was that the Irish green camouflage looked remarkably like that of the Afghan National Army. I was concerned about it, so one day I jumped out of my armoured car at the gate in Kabul and got one of my guys to take a picture of me standing beside an Afghan soldier. I sent it back to Ireland with the message, 'Spot the difference. We need desert-pattern uniforms.' While

NATO troops were walking around confidently in sand-coloured camouflage, we were still wearing our khaki-and-dark-green uniforms and it was impossible to tell the difference between Irish military personnel and the Afghan National Army. After a lot of aggravation, to put it mildly, we got new uniforms sent out to us, and they were perfect. I had some sympathy with their logic: 'Lane's making waves out there, give him what he wants,' or 'Don't give him what he wants!' This time, it really did matter, because the right uniform saved lives.

CHAPTER 13

THE RIGHT STRATEGY

SAF was suffering serious casualties from IEDs, so my job was to build up what was known as a counter-IED strategy to reduce the number of casualties. But first I had to persuade McNeill to take it seriously. One day, he said to me, 'Lane, why don't they come out and fight us like soldiers, what's with these IEDs?'

I said, 'Sir, why would they? If they stick their heads above the ground, they get hammered. And they're very effective!' It took me a long time to get through to him, until I was out one day with a British photographer and spotted exactly the example that I thought would persuade McNeill. A tank was parked at the top of the hill and further down a man was working on an IED, dressed in a full bomb suit. The next morning at the CUA, I put the photograph up. I said, 'Sir, I haven't explained myself very well, so I think this photograph does it. All our focus here is on that.' I pointed to the tank. 'S-16s, Predators, UAVs ... but 55 per cent of our casualties are coming from IEDs, so this is where we need to focus.' My briefing had a real sense of urgency to it, because we'd just lost five soldiers in Kandahar to an IED. Furthermore, the Taliban were developing increasingly clever devices: we used to use metal detectors to find IEDs, so they developed an all-wooden IED from a hollowed-out log, complete with wooden pressure plate on top, to get around that problem. It was chillingly effective.

I had gone down to investigate the incident in Kandahar, so now was my chance to make my point.

I completely understood that McNeill was a busy man and needed to see the big picture, but when he'd paid no attention to my brief, I knew that I was on a hiding to nothing. However, less than a week later, I would be sent down to the very same track to investigate another IED, which claimed another six casualties. It was time to take the gloves off. I got photographs of the people who had been killed and at my next briefing I put them up on screen, to put a face to the fatalities. I said, 'Sir, this is the war. The tank is a symmetric war, which is limited out here, and the man wearing the suit is the asymmetric war, dealing with IEDs. Fifty-five per cent of our casualties are coming from whatever is at his foot there. That's what we have to sort and, in fact, I'm not doing my job because we're losing more people than ever. I'm meant to be responsible for developing the counter-IED strategy, so I'm not doing my bloody job.'

I still wasn't getting through. McNeill simply indicated that I should shut up and sit down. That's what I should have done, of course, but I didn't! I couldn't help thinking, *Ray, you're too old for this.* In fact, the only person older than me in the room was the boss. Normally, at the end of a brief you say, 'Pending any questions, sir?' Instead, I said, 'Sir, as I know there are no questions, I'll end the brief.'

You could have heard a pin drop. I knew that I'd crossed a boundary, dressing down a four-star general in front of his staff. But to be honest, what had I got to lose at my age? I was 54 and had been all around the world, so I was too long in the tooth for classroom stuff. I wanted to go home: I was exhausted, I wasn't sleeping at night, the air was dirty, everything felt wrong. And more importantly, I also thought that dealing with IEDs was a way of saving lives.

McNeill barked, 'Lane, what do you mean by "as I know there are no questions"?'

'Sir, you were tapping your fingers as I briefed, which indicates that you've had enough of it, and I appreciate that – you didn't want to ask any questions.' He turned back to the next person and the brief continued, but I could see that he had a face like thunder. As soon as the briefing was over, he jumped up from his seat, looked at me and said, 'You. Bunker. Now.'

I slunk off to the general's room, which was housed in an underground bunker, and there he was with his staff standing around, waiting for orders. As soon as he saw me, he buried his finger into my chest, ready to let rip. He did, finishing with 'Lane, you're a great man for giving me problems, but you never give me solutions.'

'Maybe you don't listen, sir. I've been trying to give you guidance for so long and we've now reached a stage where we're losing all these people and we're still going on about F-16s and f*cking tanks.' I was reminded of Gus Pagonis's index cards and wished that I had a set on me to help me make my point – but this was a different situation.

'Well, come on then, what do you want me to do?'

'Restructure the CUA, so that there's less talk about air assets and navy assets, more talk about IEDs and how we're going to reduce and mitigate the effects they're having on our people and the local population.' Now, that was a big deal, because the CUA represents a great chance for personnel to be seen in a favourable light. After all, who doesn't want to be a four-star general? I, on the other hand, was at the end of my career, so I'd nothing to lose. I added, 'On Sunday morning, give me a half-hour where I will go through one item in particular, an IED, and how I think we can mitigate and reduce its effect on our people and the population. Just 30 minutes on a Sunday.'

'Done,' he said. 'But you know the pushback there's going to be from people about this.'

'Well, sir, that's why you're a four-star general,' I replied. I got to the door, turned and saluted him.

He looked at me and said, 'If you ever speak to me like that again, Lane, it's Iceland I'll send you back to, not Ireland.' But he had a smile on his face.

From that moment on, we knew exactly where we both stood. The next morning at the CUA, I put that slide of the soldier in bomb-disposal gear with the tank in the background and I explained what it meant to me. I quoted from a US organisation, JIEDDO, the Joint Improvised Explosive Device Defeat Organization: 'No other widely available terror weapon has more potential for media tension and strategic influence than the IED.' I wasn't suggesting that any other tactic be ditched, but that the focus now had to shift slightly. I reminded them that the Spanish had withdrawn from ISAF because they had lost so many soldiers to IEDs. All that was needed now was for another major country to pull out of the coalition of the willing, and they were in trouble.

Mission accomplished. Following the briefing, we became a lot more serious about our counter-IED strategy. I explained to them that the explosion itself was only the end of a long process.

That process could take place up to a year before with the supply of explosive material, transporting it to a bomb-builder, who would coordinate with someone who would plan the explosion, right down to the person who detonated the device. We had unbelievable assets in the ISAF, so we used them to watch people 24 hours a day. We used Predator drones to survey them and build up cases against the serious players in the field. We watched people who were developing suicide belts and suicide bombs. For example, the original suicide belts were always bulky and heavy, so we'd send out pictures and a 'flash note' or brief report, telling personnel what they looked like, how to spot one and what to do in the event of one being detonated. We also had forensics, so that we could analyse bomb parts and pick up fingerprints and other intelligence from them. We became experts at the defusing of suicide belts also.

Of course, the Taliban would get hold of the information and use it to change strategy, for example replacing the bulky suicide vests with more streamlined versions, and we'd take note of that, and so on. It was a game of cat and mouse, except that real people were being killed.

On a number of occasions, Special Forces would identify key people in the Taliban, and we'd follow them to a location, where we'd watch them, sometimes for days. By this stage, ISAF would have put a lot of effort into finding them and their bomb-making locations, but what happened next was decided by McNeill. My job was to assess the danger and report on it to him and to our legal people. McNeill would look at the drone pictures very closely, asking questions. 'There's a building there, Ray – who's in the building?'

I might reply, 'No one. It's empty.'

'Are you sure?'

'It's a hundred per cent empty.' The stakes were high: if I said it was empty and someone was in it, a missile would be heading in their direction. If I said, 'We think it's empty, but we can't be guaranteed,' his reply would be: 'Okay, then cancel the operation.' McNeill was a man of honour, and he was absolutely right, no matter how frustrating it might have been. I had seen some of the handiwork these people had done; they were killing so many people it was unbelievable. I didn't really have huge sympathy for them … but McNeill was right to want 100 per cent proof before he would authorise any retaliation. That was why he was a four-star general and not me, I suppose. It was his job to make life-or-death decisions.

◆

In spite of the volatile situation, there were a few lighter moments on that tour. Under the tutelage of my sergeant, Gerry Setright, we had perfected a routine for dealing with breakdowns and punctures, which might happen on the poor-quality roads as we went to and from various locations. As I said earlier, Gerry was an exacting taskmaster, and we all knew our roles if we got a puncture. I was the boss, so obviously I wasn't going to get my hands dirty, so Gerry told me, 'Sir, what you do is you get the big bottle of water, and you pour it over the nuts on the wheel – do you think you can do that?'

'I think I'll manage that, Gerry.'

I became very good at pouring water over the nuts while another guy loosened them, and before long, we had our routine down to

12 minutes. That's fast in an armoured vehicle. To protect ourselves from being fired on while we were changing the tyre, we were each to take a position around the vehicle. Gerry said, 'Sir, your role is, you'll go 50 metres beyond the vehicle and take up a defensive position, so you give security to the front of the vehicle.' We'd have people on the wings then and people at the back, forming a corral. I nodded, thinking that I was never going to use this information, so it needn't trouble me.

One day, we were right in the middle of Kabul when we got a puncture. Not ideal. We also happened to have an American colonel with us, which was against regulations – we were taking him to the airport in Bagram as a favour. As per Gerry's orders, we all clambered out and the American went forward 50 metres, crouching down then with his weapon. Gerry and I set to, undoing the screws.

The American looked at Gerry, then at me. 'Sir?'

'Fifty more metres,' I said. So he went another 50 metres up the road and he took up the firing position again while we finished up changing the tyre. Gerry yelled at him, 'Another 50 metres.' So, he ran on another 50 metres and crouched down again with his weapon at the ready. By this stage, he was a good bit up the road. I gave the order for everyone to get back in the vehicle, and the American came running back, huffing and puffing. We clambered back into the vehicle and drove off.

The American was silent for a bit, then said, 'By the way, Ray, can I just ask a question? Purely from an operational point of view, when you ask me to go forward 50 metres, then another 50 metres, what's the purpose of that?'

'Oh,' I deadpanned, 'it's in our procedures. If an American or German are on board and the Taliban are going to attack us, we give you to the Taliban and that protects us. They get you first, so it gives the rest of us an extra couple of seconds.'

The American didn't say another word. Gerry sat in the front, driving, trying to keep a straight face. We kept the façade up until we got to the airport. Thankfully, the American got the joke!

—◆—

So, now I had my counter-IED strategy up and running and we were developing a very good relationship with the Afghan military and the NDS, the secret service, and it was time to move onto the next stage. I decided with my team to organise a series of exercises to put into practice what we'd been talking about at the CUAs for the previous few months. The goal of the exercises would be to get the Afghan National Security Forces and NATO on the same page when dealing with an IED situation. Saoirse Nua was the name of the programme, and our byline was 'cooperation and coordination saves lives'. We prepared a presentation both in English and in Pashtun, and we briefed McNeill and other personnel on our plans. In our game, we use the expressions 'left of boom' and 'right of boom', i.e. everything that would happen before the explosive and everything that would happen after.

The only slight problem was that the British weren't entirely happy with the title Saoirse Nua. One of my British colleagues warned me, 'It has connotations, Colonel, back to Northern Ireland, that mightn't be appropriate in this case.'

However, McNeill said, 'I like it, move on,' so the deal was done.

When I had finished my presentation, McNeill sat back and thought for a minute. 'I think this is a great idea, but you do realise, Lane, that once the Taliban hear about this, you'll be targeted.'

'Well, we'll have a contingency plan, sir. I intend to make billboards, huge billboards, and put them up prominently in Kabul for the people to know when these things are happening and where.'

'Why are you doing that?'

'The bottom line is, sir, we're here to help these people, not fight them, right? Coordination and cooperation aren't just between the ISAF and NATO, it's the people. So I want the people involved.'

'You know you're making yourself a bigger target,' he reminded me.

'Leave that to me. I'll develop robust force-protection plans for our security.'

Of course, he wasn't just referring to the Taliban. Later that afternoon, I received a call to his office. When I went in, he said, 'Lane, I want to tell you this is a brilliant idea, but do you realise how many people are against you inside this place? We have our training people who think they should be doing it and are not doing it, we have other organisations that know they should be doing it and are not doing it … the thing is, they hope this fails and it'll be the end of you. If it doesn't fail, there won't be enough carriages in the train for the number of people saying it was their idea. Go and do it.'

Duly warned, I went over to the Afghan training centre outside of Kabul, looked at the ground, spent days putting the whole plan together, then I looked at security, planning for possible Taliban attacks, the whole business. I prepared yet another slide presentation with our objectives, both in Pashtun and English. These were:

▶ To understand and clarify roles and responsibilities in each organisation.

▶ To identify gaps in our capabilities.

▶ To improve individual performance.

▶ To identify future challenges.

Next, we identified areas for the exercises. These areas would stand in for different locations in Afghanistan and would present teams with IED situations to resolve. It was quite a logistical feat, as you can imagine, because we needed to find not only a suitable spot for the exercises, but also places to feed up to 500 delegates, have a control centre, security gates, a controlled parking area and so on. In the exercises themselves, we would replicate real-life incidents, such as IEDs that had exploded in Kabul, and we'd look at people's roles and responsibilities, the correct signals, the planning needed to tackle IEDs, the equipment, the forensics.

We put huge billboards up in the local language to advertise the exercises, so no one, including the Taliban, could be in any doubt that they were happening. In the end, we did three sets of exercises. We trained our people in the techniques, tactics and procedures they would need in the IED field – job done. The exercises were a great success. So much so that on the last exercise, Saoirse Nua 3, they decided to say hello by attacking one of our convoys in Kabul. Thankfully, ISAF had no casualties.

I would have been happy enough with that, but ISAF were delighted with the outcome. At the end of my seven-month stint in Afghanistan, I received a commendation from McNeill. In the accompanying report, he wrote, 'Shortly after his arrival Kabul experienced a series of horrific bomb blasts. Colonel Lane reacted quickly to this, conceiving, organising, and leading the Saoirse Nua exercise series; thus demonstrating outstanding foresight and initiative.' I'm not saying this to toot my own horn, but because of what the Saoirse Nua exercises achieved. 'As a direct result of these exercises we have seen a direct improvement in the management of IED situations in Kabul by the Kabul City police and Afghan Army. Furthermore, it has developed common training and working practices at ANSF and NATO, significantly improving the work and understanding between the ANSF and IASF.' All very impressive, you might think, but the real result was that the proportion of casualties from IEDs in Kabul went from 15% to 4%.' The template was replicated in Kandahar, which was particularly dangerous for ISAF personnel.

———◆———

I left Afghanistan on a Sunday in September 2007. It was time to give my final CUA briefing, so off I went with my slide presentation, as normal. I did the rehearsal with the German commander, standing on the X, and after a few slides, he said, 'Ray, that's fine. I hear you're leaving today?'

'I am, sir, yes,' I replied.

'Well, thank you for all your efforts.'

I nodded my thanks and we all waited for McNeill to come in. The presentation proceeded as usual, and then I stood up to give my final presentation. I was about to start, when McNeill interrupted. 'Going home to the Emerald Isle?'

'Yes, sir.'

'Continue,' he said.

Then I brought up my next slide. 'Now, I have to apologise to the German general here, but this is a slightly different presentation to the one I gave at the rehearsal.' My German commander's face paled. McNeill looked at me, delighted. 'Slide one, please,' I said to the American sergeant in charge of the presentation, and up came a slide in block capitals: HOW TO WIN THE WAR.

'Great question. I hope the next slide is going to give me the answer,' McNeill said.

'Slide two, please,' I said. The second slide appeared: GIVE THE TALIBAN POWERPOINT. There was an eruption of laughter. 'Since I've come out here seven months ago, the most used phrase has been "How many slides?",' I explained. 'In which case, we'll certainly win because they can't produce any more slides than us.' At this, even the German laughed. Then up came slide three: 'From the Irish Contingent, we wish you all the best in the future out here: cooperation and coordination save lives.'

That was it, the end of my seven-month stint. It had been worth it, I thought, as I was driven back to the airport. We had come out here and made a difference. In fact, I thought, it would seem that the ISAF/ANSF combined forces had made a difference. So, when the Taliban retook Kabul in 2021 without even a murmur, I felt sick. What had all our efforts been for, if we were simply going to run out of the place, leaving so many of the Afghans who'd helped us behind? As a gesture, I returned the shiny medal that had come with my commendation. It didn't seem appropriate to hang onto it in the circumstances.

All of this brings me back to the question of 'why', as my colonel had told me all those years before. You can do anything if you know the reason why. We all knew why we were in Afghanistan and what we wanted to achieve: to provide security and stability, to secure funding for a rebuilding programme in the country and to train the Afghan military to the right standard to protect the country. Seeing it all vanish in a couple of days of chaos was galling. I really did ask myself what on earth my forty or so years in the Armed Forces had been for. Perhaps that's a question that I alone can't answer.

This year, I was asked to go to Ukraine by UNOPS to advise on their counter mine/IED strategy. (UNOPS provides infrastructure, procurement and project management services for a sustainable world.) On the morning we left the country, I went out into the city of Kyiv for some air. I wandered down to Maidan Square, the centre of the 2014 revolution.* The square has also become a place of pilgrimage for those who have lost loved ones in the recent Russian invasion of the country. Each person's absence is marked by a Ukrainian flag and, sadly, there are more of them each day.

Beside one of the flags sat a little girl, bawling her eyes out. I went over to her and gave her a big hug and I asked her what was

---

*   There's a documentary produced by Oliver Stone on the events of 2014, *Ukraine on Fire*, well worth watching for the interview with Vladimir Putin.

wrong. In broken English, she told me that her brother had died the day before and that she was about to put the flag with his name on it in the ground. After we chatted for a while, she thanked me for the hug and I walked away, tears rolling down my face. I have met this girl's brother in every conflict I've been to and now, at the age of 70, I can say that I've seen nothing good come out of any of it. Arguably, I've seen worse conflicts, but I've been able to walk away and say to myself, 'I didn't do this. Someone else did it and I'm trying to help.' Now, I just couldn't.

In the past, if you'd asked me about mortality, I thought I'd live for ever, but now as a father and grandfather, I value life more now than I ever did, so to see it being tossed away like that is something that affects me deeply. I will never forget that inconsolable little girl in Maidan Square – perhaps that's a good thing. As Dwight D. Eisenhower said: 'I hate war as only a soldier who has lived it can, only as one who has seen its brutality, its futility, its stupidity.'

CHAPTER 14

JUSTICE IS SERVED

When I went to Bosnia in 1992, I made a decision to keep a very comprehensive diary from day one. Every night in my little office, I'd fill out the day's activities in detail, for sanity's sake, to somehow make sense of what I was seeing. But there was also the thought in my head that somewhere down the line, there might be a reckoning. The local warlords used to laugh at me when I'd tell them that I was taking notes, but the people of Bosnia had the last laugh, thankfully, with the founding of the ICTY – the International Criminal Tribunal for the former Yugoslavia – in 1993.

A full 17 years later, the ICTY called me to give evidence about what I'd witnessed in Bosnia during my time there. Perhaps you might wonder why the process takes so long. So did I, to be honest, but to try 161 individuals, to gather evidence on their crimes, to interview many traumatised witnesses and victims and to prepare a case isn't a speedy process.* Also, in the former Yugoslavia, like in so many wars, those responsible had gone to ground: one of the most barbaric criminals, Serbian leader Ratko Mladić, was only finally located – in the house of a Serbian relative – in 2011. Radovan Karadžić, leader of the Serbian political party, was living as a doctor of alternative medicine in Belgrade until his arrest in

---

* See https://www.icty.org/node/9590 for the story of the ICTY's achievements in simple form.

2008. In her piece for Reuters in July of that year, Ivana Sekularac describes him: 'In another picture released by Serbian authorities, he looks tired and bespectacled, not the robust politician charged with orchestrating the murder of 8,000 people in Srebrenica and being responsible for the death of 11,000 in the 43-month siege of Sarajevo.' The banality of evil once more.

When the ICTY came calling for me in 2007, I had diaries of material and carefully recorded notes. I was glad that I had them, because when I turned up at the Hague to give my testimony, the two men with whom I had the most contact, Prlić and Ćorić, looked so ordinary: two elderly men, their best days behind them, in too-large suits. However, their arrogance, their complete lack of contrition, was intact.

To set the scene, I had been interviewed at some length by investigators in Dublin in 2004 and was thankful for my diaries. Memories fade and it was very helpful to read them and to be able to recall specific incidents and events to add to the investigators' folders. They had all the reports that I'd sent back to HQ to my bosses, General Cranston and the Portuguese ambassador. Then, when I was out in Afghanistan, I was asked to go to the Hague. Now, I obviously wasn't on my holidays in Afghanistan, so I really needed a bit of mental clarity before I gave my testimony. Even with the help of my diaries, I had closed the door on my time in Bosnia some years before, resolving to put it in some faraway vault in my memory. To have it all brought up again was disturbing, to say the least.

However, I wanted to give evidence for those poor unfortunate victims of Ćorić, Prlić et al., none of whom were able to forget, so I went to the Hague to do my part. I was accompanied by a legal officer from the Defence Forces to offer support and guidance. Frankly, he was as overawed as me at the spectacle that awaited us. Attending a court like this is a fascinating if upsetting experience.

I arrived in the Hague on the Friday and the court was to begin on the Monday, with my testimony scheduled for Monday and Tuesday of that week. I had a weekend to refresh my memory. On the Saturday, I was taken to the ICTY HQ in the Hague and presented with a huge pile of documents, which I had to read before the Monday. While the sun shone on the buildings of the Hague and people bustled around the streets, I spent the weekend in my hotel room, going through the documentation, in particular my witness statement from 2004. After the hot sun and dust of Afghanistan, it all felt a bit surreal. I looked at the incidents I'd recalled in my witness statements, and it felt as if they'd happened in another lifetime. Frankly, I was dreading Monday.

However, Monday came, as it always does, and off we went to the court, a series of gloomy grey-brick buildings in front of an oval lake. It was hard to imagine that this building also housed those who were about to go on trial. As I stood in front of the entrance, I remembered my long-ago dealings with my Croatian adversaries, and wondered if I'd be capable of doing the victims justice.

We were taken to a small room to be briefed about the proceedings and what we could expect to see in court. It was also explained to us that we'd be sitting in court without notes, which I didn't fully understand, but went along with. The atmosphere inside the court is extremely intimidating for an ordinary citizen.

If you are a witness, you sit alone at a podium on the right-hand side of a large courtroom. On the left, in my case, were the six defendants in the case, alongside their numerous legal teams, all tapping away on computers. In front were the three judges, clad in the blue of the UN. Behind me, protected by a wall of glass, were the relatives of those who had been victims of anything from mass rape to genocide. Because of my job, I'm used to being in control of situations, and here I most definitely wasn't, and I felt nervous and out of sorts.

Judge Jean-Claude Antonetti was the presiding judge; he bustled in along with his colleagues and went through procedures for the day. He introduced me and thanked me for coming, and then he handed me over to the defence attorney for Prlić, one Michael G. Karnavas. Karnavas was a capable and agile thinker, not to mention relentless. When my legal colleague saw him, he said, 'Oh my God, Ray, this is the A-Team.' It sure was, at a rumoured $5,000 a day for Karnavas's services alone. I have no idea where Prlić got the money, but I suspect it wasn't out of his own life savings.

Karnavas came at me like a machine gun. He did not give me a moment, firing question after question at me about my recollection of events, questioning my recall, digging around incidents and implying that I'd misremembered them. 'So, Mr Lane,' he'd say, 'you don't recollect that event.'

'Not at this point in time,' I'd reply nervously, all memory having been wiped out in the tense atmosphere of the courtroom – and because I was without my notes.

'So, you don't recollect it then,' he would say crisply. 'Moving on ...'

This process started at ten with a short break for lunch, when I was sent back to my little room, alone. I then returned for more legal battering until six o'clock that evening. I was wrung out at that stage, but Karnavas was fresh as a daisy, or so it seemed, as he jumped to attention and said to the judge, 'First of all, I'd like to thank Mr Lane for coming here from Afghanistan. We appreciate that it's not easy. We understand that.' Then he added, 'But Judge, I asked Mr Lane three key questions. And I failed to get three clear answers. So, the question is, do we need to reconvene on Tuesday? Could we not just stop today?'

To be honest, I was sitting there thinking to myself, *I can't take another day of this*, but then I corrected myself. *Come on, you're better than that and you owe it to all those people you worked with in Bosnia.* I held my breath while they deliberated for a good while. Eventually, they announced that we would be resuming the following day at 0900 hours. I had been given another chance.

Back in the darkened car which drove us back to the hotel, my colleague said to me, 'You know, Ray, if that was a soccer match, you lost that seven-nil.'

He was right. I'd been rolled over, made to feel like the guilty party. Still, I couldn't help snapping at the poor guy, 'Any more smart comments like that, just keep them to your f*cking self, right?' I wasn't angry with him, but with myself.

Back in the hotel, I brooded over the day's proceedings before deciding to go for a run to clear my head. Running is my way of processing things, and it had saved me after Bosnia, so I went out the back of the hotel and I'd say I ran the marathon. I started running at seven in the evening and I was still going at eleven that night, up and down up and down the hotel garden, Karnavas's questions turning in my head. Slowly but surely my brain started working.

I got back to the hotel, had a shower and sat in my hotel room until three in the morning, writing and memorising notes for the following day.

I was a different person on Tuesday. When I was called in from my little cell, I asked the judges if I could address the court. 'I'd like to refer to the questions I failed to answer yesterday, and I'd like sufficient time to deal with them today.'

With that, Karnavas jumped up in his seat, robes flowing. 'Your Honours, he had his opportunity yesterday and he failed to take it.'

I looked at him and the words came out of my mouth: 'Counsel, sit down.' To my surprise, he sat down meekly.

Judge Antonetti said, 'You take as long as you want today, Mr Lane, to answer those questions.'

So, I took the whole day. I answered the three questions put to me by Karnavas, and I also managed to take a lump out of two Croat leaders, Praljak and my old friend Valentin Ćorić. To me, the most dangerous thing in the world is an army officer with a map. I was giving evidence about attacks on villages using heavy artillery and Praljak's team produced detailed maps of the area, saying that the weapons they'd used hadn't got the range to get to those villages. He had measurements and distances to prove it.

I said, 'Mr Praljak, you were using supercharge and that gives you the extra distance.' (Supercharge is ammunition that will give the person firing the extra distance they need to hit a target.) As it happened, Praljak, who in his wisdom had decided to cross-examine me, didn't know what supercharge was, so his legal team persuaded him to sit down before he did any more damage.

And then we got to Ćorić. I took a good, long look at him, a man who in any other circumstances might have been a friend. I'd developed a good relationship with him all those years before and I'd got to know his wife and family. But I'd also told him that if there was ever a day of destiny, I would be giving evidence against him, no matter what kind of relationship we had.

When it came time for his team to cross-examine me, they simply said to the judge, 'No questions.' I was a bit taken aback, but when I was leaving the court, Ćorić came down and shook my hand and said to me, 'You told us that there would be a day of judgement and I can't argue with anything you said.' Ćorić took his 16-year sentence without protest. But lest you think that I admired the guy, I was well aware of what he'd done. They were all caught up in evil in Bosnia, but I knew that as a member of the military police, he had committed more crimes than most. The fact that he took his sentence on the chin was the very least he could do for the many victims.

Sadly, this didn't apply to other members of the Croat forces. In one of many terrible incidents recalled at the court, two brothers, Milan and Sredoje Lukić, had been charged with crimes against humanity, or more precisely, 'Persecutions on political, racial and religious grounds; murder, inhumane acts and extermination.' (From the ICTY case information sheet, Višegrad.)

On one occasion, the two brothers rounded up 70 people and locked them in a house. They put petrol on the carpet of the house and then threw an explosive device in to set fire to it. Anyone who managed to escape was shot as they came out of the front door. Their defence challenged the very occurrence of the fire; they got experts in, and they said that even if it did happen, the explosive that they used wouldn't have had that impact on the house. They were lying, of course, and as technical witnesses, our Ordnance School proved it. But what struck me more than anything was the brothers' attitude to one witness who had survived the massacre and stood up to give her testimony. Like the school bullies that they probably once were, they covered their mouths with their hands and proceeded to laugh at her and make obscene gestures when the judges weren't looking. I said to myself, *Merciful hour, she's been through this, and this is the reaction to it?* When it came time to give my technical evidence, I made reference to this behaviour to the judges.

Why did they behave like this? God only knows, but it was another reminder of how the ethnic conflict of the Bosnian War had totally stripped so many people of their humanity, had turned them into monsters, and of how adherence to a religious belief can blind people to their fellow human beings. Thankfully, they got their just deserts: life and 30 years in prison respectively.

At the end of the second day, we were driving back to our hotel, and I said to my legal counsel, 'Well, how did you think the match went today?'

'Oh, you beat him seven-nil today for sure.'

I was scheduled to be available until that Friday evening, when the court closed, and the judges thanked me for my evidence. I went back into the little cubicle to find two people sitting there, waiting. They introduced themselves as a medical doctor and a psychiatrist, and for a second, I wondered if they were coming to take me away.

The psychiatrist said to me, 'Well, Mr Lane, that's been a very difficult week. How are you feeling?'

'Good now, no problem,' I replied.

'Great – can you explain then why your coffee is all over your shoe?'

I had been standing there talking to him and hadn't realised that my hands were shaking so much that I was spilling my cup of coffee all over my shoes.

'Sit down,' he said gently.

I lowered myself into the chair and my whole body started to shake. That had never happened to me in all my years in the Defence Forces. The two doctors settled me down, and I have to say, they were superb at talking me back down from wherever it was I'd gone. I'd thought I was doing just fine, but my body was telling me something different. Trauma is a word we bandy about nowadays, but I could feel it in myself as the memories of what had happened in Bosnia came flooding back to me.

Still shaken, I returned to the hotel, before my legal counsel and I decided to head out for a meal. 'How are you feeling?' he asked me as we sat down.

'Ah, not a hundred per cent, but a bit better. Listen, I don't want to talk about the last few days. Let's talk about anything else, but not Bosnia,' I said. Almost as if he'd heard me, who walks into the restaurant only Michael Karnavas. Before I could say anything, my counsel, seeing an opportunity to talk to one of the biggest names in the legal business, called him over. I absolutely did not want this – not after the beating I'd taken on the Monday. However, politeness prevailed and Karnavas and his companion, a Croat, sat down with us. We made small talk for a bit, then Karnavas said, 'I want to say, Mr Lane, that anyone who survived Monday, really, has some mettle.'

I thanked him politely, hoping he'd let it drop, but he continued. 'Let me tell you our strategy. I can tell you now, because it's over. We knew you were in Afghanistan. We knew you wouldn't have time to prepare the way you'd want to, so the idea was that we would hammer you on Monday and then there would be no need to bring you back. It'd all be over. Then you said to me on Tuesday morning, "Sit down, Counsel," do you remember?'

'I'll never forget it.'

'What did I do after I sat down?' he said.

'Ehm, you went on your computer.'

'Do you know what the message was? "It's over."'

When I came home from the ICTY, I don't think I slept for a month. Memories of the terrible things I'd seen and heard came right back up to the surface, particularly at night, when I'd lie awake and go over the past. I'd remember snatched conversations and terrible images of what I'd seen: the poster that had appeared all over Mostar, showing me embracing a Muslim, the two of us standing in a pool of Croatian blood; the removal of the entire Jewish population of the town, forcing them into Montenegro, shooting any who didn't comply. The sinister 'messages' I received in the form of a bullet hole in the windscreen of my car, a booby-trap device lain on the bridge at Mostar … At no time in my life had I experienced fear like it, and I'd seen just about everything, but when I went into that little room on the Friday after my testimony was complete, I'd completely lost it.

Maybe Karnavas's behaviour in the court had been a trigger, I reflected. I'd watched him strut around the court for a few days, his top-class brain, his court presence used to defend the likes of the Lukić brothers. To him, the court was a stage and he was giving the performance of a lifetime.

But, I reasoned, I'd been able to turn it around the following day. Fuelled by anger, yes, but also a feeling that I needed to stand up for myself, for Volmer, for Alenka and all of those who had worked so hard in Bosnia, and for the innocent victims of the war. But at what cost to me?

I had a conversation with a friend recently about buried memories, and I'm convinced that this process is essential in helping us to deal with terrible things. Maybe those who came back from the First and Second World Wars had the right idea in never speaking of it again, to spare themselves and their families. I know that when I came home, I needed to put these memories away: if I'd let them get the better of me, I don't know where I'd have got to.

Thankfully, the normality imposed by Bridget, who was a bloody good nurse in a hospital, and our two children, returned me to the land of the living. She was totally sympathetic to me, but she also let me know that our life at home was important too. That life goes on. That, and my go-to meditation of running, restored me to some form of normality.

Having said that, people often ask me if I think that war crimes tribunals and other post-conflict investigations achieve anything other than dragging things back up that should remain buried. That's a very good question. In 2009, I provided technical evidence to a UN fact-finding mission on a conflict in Gaza which had taken place during a three-week period in 2008. Public hearings took place in Geneva under the umbrella of the Goldstone Inquiry, called after Justice Richard Goldstone, from South Africa, who had been appointed head of the mission. Also appointed to the mission were human rights law experts Hina Jilani, Christine Chinkin and Colonel Des Travers (retd) of the Irish Defence Forces, someone I know well. The team in the Ordnance School provided two hundred or so pages of information on the types of weapons that had been used by the IDF and by the Palestinian factions, notably Hamas. We prepared a detailed dossier of the ammunition used, the bombs employed, the rockets sent into Israel by Hamas, and the use of white phosphorus in IDF weapons. The fact-finding mission's goal was as the title suggests: to find out exactly what had gone on in Gaza during those weeks and who was responsible, and to find out if any war crimes could be said to have been committed. The conclusion of the report was that potential war crimes and crimes against humanity had been committed by both sides and that both sides needed to conduct detailed investigations of their conduct during the conflict.

Nothing controversial there, except that upon publication of the Goldstone Report, a storm erupted, during which the conclusions of the report were attacked with such vigour that Richard

Goldstone later withdrew his support of the report. In a piece in the *Washington Post* in 2011, he appeared to reassess his earlier conclusion that civilians in Gaza had been intentionally targeted by Israel: 'I continue to believe in the cause of establishing and applying international law to protracted and deadly conflicts. Our report has led to numerous "lessons learned" and policy changes, including the adoption of new Israel Defense Forces procedures for protecting civilians in cases of urban warfare and limiting the use of white phosphorus in civilian areas. The Palestinian Authority established an independent inquiry into our allegations of human rights abuses – assassinations, torture and illegal detentions – perpetrated by Fatah in the West Bank, especially against members of Hamas. Most of those allegations were confirmed by this inquiry. Regrettably, there has been no effort by Hamas in Gaza to investigate the allegations of its war crimes and possible crimes against humanity.'

Goldstone's colleagues responded in the *Guardian* newspaper that 'in the case of the Gaza conflict, we believe that both parties held responsible in this respect, have yet to establish a convincing basis for any claims that contradict the findings of the mission's report'.

Had there been political interference? It's certainly true that Justice Goldstone came under extraordinary pressure to remove his name, and with Goldstone's withdrawal from the process, the report did not carry the weight that it might have done. Instead, a barrage of allegation and counter-allegation took place and there were calls for the report to be withdrawn entirely. A lot of investigative work and the testimony of those affected was critically undermined, something that will, I fear, influence any future processes in this area. It's about equity: the balanced and equitable application of the rules of international law.

But what I saw in the Hague gave me hope that even though the process isn't perfect, at least it exists. Those who had suffered were able to tell their stories. And the bad boys of Bosnia got their due. They got their sentences. But they also got the kind of humane treatment denied to their victims, which makes me very angry. And, as is often the way with these things, some of them escaped completely. I know plenty of people, whom I won't name, who are free in Bosnia now having never been charged with anything. It's small consolation that when they went to jail in the Hague to serve their time, Muslims, Croats and Serbs were all mixed in together. What an irony.

But I suppose that's what happens after every conflict, from the Second World War to Cambodia to Rwanda. Some manage to escape the net and never get the punishment they deserve. It is not a perfect system, by any means, but it beats the alternative, which is that we learn nothing and simply repeat the cycle all over again, knowing that there will be no consequences.

CHAPTER 15

AN ARMY MARCHES ON ITS STOMACH

Southern Lebanon, 1996

knew from my posting in 1988 that the two most important things to a UN peacekeeping mission are communications and food. Soldiers rely on both, for reasons of morale apart from anything. Way back in 1988, we formed a close-knit community, thanks to our experiences in those isolated outposts along the Lebanese border. We were all in it together, which mattered a lot, because communications with our loved ones back home were very difficult. There was no email, no WhatsApp, no FaceTime or mobile phones, and communication was rudimentary.

On my first UNIFIL trip, the adjutant of the battalion was keeping count of the letters each soldier was getting: if he wasn't getting any, you'd be worried. If he was getting too many, you'd be worried. It so happened that Lieutenant Lane, R., was getting less than anybody. It didn't bother me unduly. Bridget was working as a nurse, as well as looking after our two children and her mother at this point, so she rarely had time to put pen to paper. One particular Sunday, the bag of post came in from Ireland to great excitement. I didn't take much notice, until I heard my name being called out as the first letter was pulled out of the bag. Surprised, I went to get it, to find that the second was also addressed to Lieutenant Lane, R., and the third, and the fourth. It transpired that all 750 of the lads in the battalion had written a letter to me. It pays to be one of the gang and I really did appreciate it.

I would get cassette tapes from Ireland, though, which I'd insert into my tape recorder and hear Bridget's voice come out of the machine almost as if she was there in the room. One day I slotted the tape in and Bridget greeted me. 'Hello, Ray, Bridget here; all well at home, the kids want to say hello and your father's here.' Then my son, David, said something in the background about the car breaking down. It was indistinct, though, and no matter how many times I replayed it, I couldn't hear what he was saying.

Then I had an idea. I brought it up to the signal lads, who were the experts in intercepting communications and had all the right equipment. When it was replayed, I could hear David say, loud and clear, 'Mammy crashed the car.'

I'd better find out what happened, I thought. At the time, we had to put in a request to ring Ireland, which would then be booked with Bridget to make sure she was there. We'd have to go up to the communications centre at a particular time and talk to each other on a crackly radio. Not ideal.

Anyway, I got the ball rolling. 'Hello, Bridget, over.'

'Hello, Ray,' she replied. 'How are you, over?'

'What happened to the car, over?'

'How did you know, over?'

'I heard David on the tape, over.'

There was a long silence, before she said, 'You got the car serviced before you left Ireland, over. You forgot to put the cap on the oil top, OVER. All the oil went out, OVER, and I was in Roscrea with the kids and the car came to a halt. OVER. IT'S AT THE MECHANIC NOW, OVER, AND IT'LL TAKE WEEKS TO REPAIR.'

So it was all my fault. All I could do was laugh. I said, 'Everybody okay?' and she said, 'Grand,' and that was it.

Being apart from our families took some getting used to, but thankfully, we always had food to distract us. We had an extremely good chef on that first tour, a lovely guy, and he looked after me and my new team of bomb-disposal officers very well (we had been given EOD responsibility after the deaths of the three soldiers). Part of our job was doing the baggage for soldiers coming home: we would get the bags, weigh them and put them on the jumbo jet. We only got a certain cubic capacity, so we had to be very careful, but we always tried to look after people who'd looked after us. So, I popped into the kitchen one day and promised to look after the chef when he was returning home.

'Well, actually, sir,' he replied, 'there was always one thing I'd love to have done.'

'What's that?'

'I'd love to go out with you when you're blowing something up, just to see how you do it.'

I was heading out to a place called Haddatha and there was a simple mortar shell on the ground, which presented no danger, so I invited my friend along. I stress that this was not an IED, just a mortar shell. However, on the hill above Haddatha were the South Lebanese Army, basically surrogates of the Israel Defense Forces. Both forces are always nervous about potential attacks on their compounds from armed elements, but they'd been informed of our plans, so I felt reasonably reassured. I got permission from UNIFIL to blow the mortar up and off the two of us went.

I talked my friend through the procedure. 'Now, this is how we do it, right? There's the mortar bomb there. I work out the size, I work out the type of explosive. I make sure that all the people around here know what we're going to do, which they do, and that they've all been evacuated.'

He said, 'This is fascinating stuff, sir.'

I went back. I made a charge, and we came down to place it onto the device. 'When we've done it, I'll bring you down to see the hole in the ground, right?' I explained. As per procedure, I got onto UNIFIL HQ: 'Permission to blow?'

The answer came back on the radio. 'Permission to blow.'

I blew the shell up and we both walked back to see the hole left behind. He looked at it in admiration. 'That's amazing,' he began. Suddenly, the ground erupted beside us. Even though they knew what we were doing, ever nervous, the boys up on the hill had started firing on us.

'Let's get out of here,' I said, and we went running up the hill and got behind a small wall. I could hear one of their tanks moving and thought we were done for.

'Jesus, sir, this is some birthday,' my friend exclaimed, delighted with himself. Thankfully, he didn't seem to recognise the danger we were in.

Suddenly, the radio in my vehicle began to crackle. A voice said, 'Rickshaw, can you confirm, is it LMG or HMG fire?' Rickshaw was my radio name, to preserve anonymity, and the caller was referring to light machine gun or heavy machine gun fire.

I crawled out from my hiding place, grabbed the handset and yelled, 'Jim, listen to this, what do you think?' and I held the handset up so my boss, Jim Mortell, could hear the heavy machine gun fire for himself. I threw the handset back and sat back under the wall, waiting for the tank to reverse back up the hill. With a sigh of relief, I pulled the chef into the vehicle and off we both went, with him exclaiming all the way about the excitement of it all.

When I got back to base Jim Mortell said to me, 'Well, what about your voice procedure?'

I said, 'What about the f*ckin' HMG, Jim?'

He said, 'But you used my name.'

'But we were being shot at!'

———◆———

Almost ten years after this first tour, no one was more surprised than me to be appointed food officer for the 7,000 UNIFIL troops stationed in southern Lebanon. To this day, I think the job was supposed to go to my brother, Philip, who was in the Transport Corps. At the time, I was stationed in McKee Barracks in Dublin, and the phone rang off the hook with calls from colleagues asking me how I'd managed to swing that gig! I wasn't complaining. After 15 years at the coalface along the Border and that difficult previous posting, the prospect of 12 months coordinating UNIFIL's food requirements felt like something of a holiday.

As an added bonus, I could bring my family along. As anyone who has been in the military will tell you, the job puts a strain on family life, so to have Bridget, Clare and David with me would be all the more special. To celebrate, I bought a brand-new Peugeot 406 and had it shipped by container to Israel. I was thrilled with myself and so was Clare, because she loved the ad for the car, in which various people did heroic things to the tune of 'Search for the Hero'!

This time, the force was drawn from Ireland, Finland, Fiji, Italy, Norway, Ghana, France and Poland. So, my job was, in a nutshell, to ensure that there was enough food to feed these people, taking into account their various tastes and cultures, and to do it all on $6.22 a day per soldier. It seemed simple enough. I should have known it would be more complicated when I went to my office on the first day for my handover session with my predecessor and he said, 'Thanks be to God I'm out of here.' I couldn't see the problem – I had enormous storehouses full of food, freezers full of frozen goods and the prospect of trips in Lebanon and Israel to sort fresh fruit and veg and dairy. I was in my element.

The problems started with wine. The Italian and French soldiers were entitled to a wine allowance each day, believe it or not. The index cards on which we wrote the amounts by hand told me that we had 15,000 litres of white and 20,000 litres of red in stock. *Excellent*, I thought, making my way down to the wine stores to have a look. The stores were being managed by a platoon from one of the UNIFIL countries – I won't say which – and I double-checked the figure with them. Yes, they told me, it was accurate.

'Great, put it all on the ground so I can count it.'

There was a long silence, after which they reluctantly laid the bottles out. I counted them all and found that we were down 3,000 litres of red and about 4,000 litres of white. I looked at them all and said, 'I'll be back tomorrow morning and I'll do exactly the same, and I presume the figures will match the figures on this card for every product, because I'm going to go through each one of them.'

I came back the next morning and it was all there. Later, I was told that there had been cars driving around Lebanon all night getting replacements! This might seem amusing, but to be honest, I was shocked at the idea that they'd be selling our wine stocks on for a profit, because at the end of the day, this is taxpayers' money. So, I took advice and changed the structure of my stores, recruiting a superb sergeant called Marius to manage them. I had no problems with theft from then on.

Now, in 1997, I was in heaven, compared to that first trip with the Irish Battalion in the mountains almost ten years earlier. I was down on the Mediterranean coast in Naqoura, and I had my own cabin, and I was president of the officers' club. At night, I would either go running or play badminton. I started a badminton club for all the nations, so we played three nights a week. I'd be able to wander into Israel when I felt like it, before Bridget and the kids joined me. Frankly, it was a picnic compared to previous jobs.

Until I received an email from New York, and it said the following: 'The costing per soldier per day of $6.22, with effect from three months from now, will be reduced to $3.11 per day'.

I couldn't believe it. I had to feed all of these people, plus Lebanese contractors and staff, who were understood to be part of the package, on half of what I had been spending. I had to look at tightening everywhere, but no matter how much I tightened up, I couldn't bring it down to $3.11. Well, I could, but the quality of the food would suffer. And so it did. The supplier changed and a few weeks later, 17 containers arrived with food from our new supplier. I rejected 15 of them, so bad was the quality. I had lifted pack after pack of inedible food out of the containers and knew that I couldn't possibly feed it to the guys – which only left two. Murmurings about the quality of the food began to reach my ears, and I knew that if I didn't sort it out, we'd all be in trouble. What is it they say about an army marching on its stomach?

So, off I went to my boss, a very nice general who absolutely did not want me to rattle the cage. The conditions for a general in the UN are very good, so if you have any sense, you try to extend your stay. I knew that he knew that I knew that any complaints could put this in jeopardy. I wasn't surprised when his response to my complaints was less than ecstatic. 'So, Lane, what are you going to do?'

'Well, sir, I can go to Beirut and buy coffee and breakfast materials in the supermarket, but I've costed it and it's going to go from $6.22 to $17.50 a day buying it like that.' There was a method to my madness: firstly, I needed to have three months' reserves in stock, and secondly, I wanted to make the problem so big that the UN would have to change its mind.

He said, 'Do it,' so off I went to Beirut, and we bought tonnes of coffee and tonnes of cereal and bread and so on and all was well. Until the bills went to UN HQ in New York and all hell broke loose. And the soldiers weren't happy either because of the poor quality of the shipped food. It seemed that I couldn't win.

The boss called me in and said, 'Ray, you're causing turmoil here. I want you to go to New York and I want you to straighten things out.' So, that's what I did – and let's just say that the meeting was unsatisfactory and was rapidly followed by an audit of my supplies by the UN in New York. *That'll teach you a lesson* was the message, I think.

Solving the problem took an intervention from several other UN Forces, particularly UNDOF, where the support of General Dave Stapleton (the Irish Force Commander and a future – outstanding – Chief of Staff) was the turning point. Once he was shown proof of the poor quality of the food and was told that his troops were receiving rejected food, his written response with pictures to the UN in New York killed the contract. Eventually, order was restored. In fact, the chief administration officer for UNIFIL, Mr Mitchell, asked me to extend for another few months. I was happy to do so, because once the food was sorted out again and the complaints stopped, things were beginning to settle down. I now had a lovely apartment in Israel, as families weren't allowed to live in Lebanon then, and my very first visitors were my father and stepmother, Betty.

I was thrilled and surprised that Daddy was coming to Lebanon, because he'd always believed that it was full of terrorists and little else. He'd watched the civil war unfold in the 1980s, so he was wary. However, a very good friend of Bridget's and mine, Jim Cahill, was a captain in the Irish Battalion area in the mountains and he invited Daddy and Betty to see the setup there. Now, Daddy had been to the lovely seaside setting of Naqoura, so he relaxed a bit as we made our way up to the Irish Battalion area in the white UN jeep. We had a beautiful meal there and Daddy was most impressed with the work being done. As far as Daddy was concerned, I was doing nothing in HQ other than swanning around, and these were the real soldiers up here.

His judgement was confirmed when a big white armoured car (called a Sisu) pulled in, and Jim went off to get his helmet and his weapon, off to do some real soldiering. Daddy was ecstatic. The next minute, the lads got two orange crates and placed them at the back of the armoured car. My father looked at them and said, 'What are these for, Jim?'

'For Betty, to get up into the armoured car. We're going to take you along the border.'

My father was over the moon as he and Betty and yours truly clambered into the armoured car, ready for our trip. Daddy looked at the driver and said, 'What age are you?'

'I'm 19,' he replied.

'And you're driving an armoured car like this? In Lebanon. Incredible.' Jim took us around to the top of a hill where we could look over Israel and Lebanon, and we stopped off at a number of positions to talk to the soldiers. Everywhere we went, my father said the same thing to them. 'Lads, what's the food like here?' And of course, soldiers will always complain about food. 'We're not getting enough sausages, we're not getting enough of this, we're not getting enough of that,' so as they'd complain, Daddy would say, 'Ray, are you taking notes here? By the way, my son here is the UNIFIL food officer.'

It was a special day, and then on our way home, I decided to pop in and see the Fijian battalion, whom I was very fond of. I rang the battalion commander and I said, 'I'm going to call in, if you don't mind, with my father and stepmother, just to say hello to you.'

'Oh,' he said, 'we'd be honoured.' So when we drove into the Fiji HQ there were 10 soldiers in a line in a guard of honour for my father. Because he was an older person, they looked up to him and stood to attention when he walked along, inspecting them. He was in his element!

We went in for something to eat and the battalion commander was complimenting me on the food and all the rest of it. 'In fact, Mr Lane, normally, we have to buy kava, you know, the Fijian drink? And your son has been supplying that to us and we're so happy about that. It's raised the morale of my soldiers no end, it's wonderful!'

I just said, 'Thanks very much. Nice to hear something good!'

We drove back across the border to Israel, and we were sitting looking out at the Med and my father was deep in contemplation about all of this. Finally, he said, 'That was one hell of a day, wasn't it?'

'Yeah, it's an amazing country Lebanon, isn't it?'

'Amazing,' he agreed. 'I've learned so much today.' I was delighted that I'd been able to share a part of my working life with my father and to change his mind about Lebanon.

Things were ticking along nicely until it came time to renegotiate the wine contracts. I really should have left well alone and stuck to the well-trodden path of the usual French, Italian and Lebanese wines, but I decided to rock the boat and to go global. I would get wines from all over the world and we'd have a tasting day, I decided. All the wines would be sampled and at the end of the day, they'd be ranked in order from one to five. Of course, no sooner did I say this than the Force Commander came to me and he said, 'I know you're having this day and I support it; however, Ray, it has to be Lebanese wine number one, that's for sure.' I kept my counsel. I knew that any of the top five candidates would get a really good order, so I was relaxed enough about it.

The next to visit me was my immediate boss, who was French. 'I know you're having this great day and I support it, but you know, the number one wine has to be French.' I had started a diplomatic war. Undeterred, I got samples from South Africa, Italy, France, Lebanon … we even went to South America. I invited all the generals, my Force Commander, civilians and so on to the Italair building for the tasting. The Italians have the contract to fly helicopters in Lebanon and their building is as elegant as you'd expect, an expanse of white stone and glass.

So, picture the scene with us all outside gathered around a table laden with food and wine samples. I'd numbered them all carefully but made sure that the identity of each wine was hidden. As I introduced the day I could feel the eyes of my French boss upon me, and the unspoken words that it had to be French number one, and the Force Commander's, with the understanding that the first choice had to be Lebanese.

I was determined that it be a blind tasting, so uncorked the bottles of wine. The only wine that wasn't there, however, was the Italian one. I thought, *what's going on here?* The next minute, we heard the sound of an Italian helicopter flying just above us. The door of the machine opened, and a little basket was lowered on a rope containing the wine. Typical Italian flair.

They didn't get any extra marks for the presentation, but a great time was had by all. Then it came time to announce the results of the blind tasting. Italy was number one on both white and red, then the Lebanese with Chateau Kafiyra, at number two, which was a lovely wine. The South Africans came in at number three and the French came in at four. I knew there would be war, to put it mildly. Nonetheless, I went up to the Force Commander with the results. He looked at me and said, 'Take that away and put Lebanese number one.'

'Sir, we need to think about this,' I protested. 'The Lebanese are at number two and they're going to get a good, sizeable contract out of it ... and the Italians ...' I hesitated. 'The only problem I can see on this list is the French. There'll be war over this.'

He brightened visibly. 'Yes, this'll upset the French, won't it?'

'Jesus. To put it mildly.'

'Okay, leave it as it is.'

He marched over to the lectern and made his little speech about what a wonderful day it had been and thanked everyone involved. 'And now, for the results, which I support entirely. Number one, Italy. Well done to Italy, even without the helicopter. Second, Lebanon. It's wonderful for the people of Lebanon to see that we hold their wine in such great esteem. And third, South Africa, who are not part of UNIFIL, but you do produce great wines.' And he sat down to a round of applause. I breathed a sigh of relief. I'd got away with it.

The boss of the French forces waited until I was back in my office to tackle me. He came in through the door like a bull. He was so angry, ranting and raving about the unfairness of it all. I knew that I had to remain calm. 'Look, sir,' I said, 'I had no hand, act, nor part in the selection. I just ran the thing. But can I suggest that if you feel that strongly about it, you go to the Force Commander?' I didn't think he'd be foolish enough to actually do so, but he did. And he came away with a flea in his ear and a threat to send him home.

When he came back to see me, he was more conciliatory, apologising for his earlier tantrum. I said, 'Look, sir, it's all irrelevant. You fly your own food in and your own wine anyway.'

He shook his head, 'No, it's the principle of this, Ray. The principle.' Despite my reassurances to the contrary, he felt the slight. And that is the UN, in a nutshell. The stakes are deadly serious for those who serve, so perhaps this is why things that might seem to be trivial can quickly get out of hand.

When my replacement arrived out the following year, I was astonished to find that he was in his last year in the army before retirement. *God almighty,* I thought, *why would they do that to the poor guy?* Anyway, he arrived into my office and I said, 'How are you on computers?'

'Oh, great, I've done the army course.'

'Okay then, let me show you how to make an order,' I began, clicking on the relevant tabs in the spreadsheet in front of us both. 'I'm going to start now with frozen food. I have all my series of meat cuts here. I have all my people and I know what weight each man's entitled to and the costing … are you happy with all that?'

'Yes,' he said blankly. I decided to demonstrate, so clicked through pages one and two, and then I said, 'Tell you what, you sit here now and do pages three and four.'

He looked puzzled, then said, 'Ray, remember at the very beginning, you pressed some button there.'

'The on/off button?'

'Yeah. I lost you after that.'

—◆—

The deaths of the three soldiers affect me to this day. I'm sure it affected all of us who were part of that tour. No doubt it changed the lives of the families of those killed for ever.

However, the peacekeeping mission in Lebanon was, and still is, incredibly complex. The Irish peacekeeping mission is still holding the Blue Line on the border between Lebanon and Israel while war rages in Gaza. As Sky News journalist John Sparks said, 'These posts now find themselves situated at the heart of the battle zone with hostile fire from both sides landing perilously close.' In fact, RTÉ's Justin McCarthy reported on this same tour that '37 rockets were fired from a site just 950 metres away from Camp Shamrock directed towards Israel … Later, Israeli warplanes struck targets near a village close to the Blue Line …'

There's no doubting that the Irish peacekeepers are caught in the middle of a dangerous escalation in tensions, as relationships between the two sides are more fractured than ever. As battalion commander Cathal Keohane said to Sky News in November 2023, 'A peacekeeping force goes in when both parties are seeking peace and you are there to monitor, report and provide an impartial witness to what is going on. Then as now, both parties – the IDF and Hezbollah – need to actively seek peace and the prospect of that peace seems as perilous as ever.' In spite of death and injury, the Irish peacekeepers continue to do good work in Lebanon, and we should be proud of them all.

CHAPTER 16

THE HYBRID THREAT

n October 2001, I was called out to a school in south Co. Dublin, where I was told a suspect letter bomb had been found. Note the use of the word 'bomb'. I assembled my team and went through what we knew about the device as we packed our equipment and got on the road. When I arrived at the school, a chief superintendent of An Garda Síochána (known to us as AGS) was waiting for me, so I knew something was up.

'Commandant, thank God you're here. The secretary in the school opened an envelope this morning and white powder fell out onto her lap. She's very upset, as are the staff and the pupils in the school.'

'Okay, but where's the bomb? I'm here in aid of the civil power to deal with a bomb?'

He was a bit taken aback. 'I thought you dealt with this type of thing as well.'

I said, 'We do, but if I'd known this before I'd left the base, I'd have had more equipment on board.' Still, I suited up as best I could into my CBRN defence suit, complete with respirator and filter, all the time trying not to frighten the daylights out of the staff. The probability was that it was a prank; I knew that, but could I say it for sure, I wondered, as I walked into the office and spoke to the secretary.

'Have you left this office to go anywhere else?'

'Yes,' she answered. It turned out that she'd gone to the kitchen to make a cup of tea.

It's beginning to escalate, I thought, as I asked her where.

I went back to talk to the Chief Super, who informed me that the children of the Israeli ambassador attended the school. Now, the threat analysis was changing. What had started as a possible hoax had now grown into a significant incident. Fundamentally, I thought that the white powdery substance wasn't harmful, but if you're not sure, you have to prepare for the worst. What if it did turn out to be anthrax spores? You can't ignore that possibility. So, we called in the fire brigade because they also had a decontamination kit, and I had all the available specialist Defence Forces equipment sent to the location along with our second EOD standby team. What had started with our two vehicles at the gate of the school had now turned into the scene of a major incident.

Then, the Chief Super received a call on his phone from the Assistant Commissioner. One of the kids in the school had rung the *Gerry Ryan Show* on RTÉ 2FM and was now giving a running commentary on the whole incident, including what everyone was doing. I got my binoculars out and I scanned the school windows until I spotted a student with a phone. It wasn't entirely unexpected and frankly no big deal, but we had to take it off him and that was the end of Gerry Ryan.

In the end, we had to set up a mini-village for triage and every person in that school went through shower decontamination, reclothing in specialist kit and a transfer by ambulance to St Vincent's Hospital for blood tests. It was an enormous job, involving hundreds of staff and students. Finally, at eight o'clock that night, we left the scene and headed back to Clancy Barracks. On arrival, we got a phone call. White powder had been found in Dublin

Airport. So, off we went, kitted out with our full decontamination kit, to begin the process all over again.

This time, as soon as I entered the room where the powder had been found, I had a hunch. Still, just like I had in the school, I interviewed the people present, and someone dropped a little bit of information that made it apparent to me that this was a hoax. It turned out that an employee had wanted to go home early and figured that an anthrax scare was the ideal way to do it.

Now, I could have insisted that we go through the whole incident-management protocol again, but using my common sense, this situation was over in five minutes. However, what is significant about both of these incidents is that they represented a potentially new and much more complex threat than anything we had seen before.

About a month earlier, after the terrible attacks on the World Trade Center, a series of letters were sent to US public figures, including senators and journalists. The wording of the letters was clumsy, but what was significant was what they contained: a lethal weaponised bacterium called anthrax. Anthrax is a biological weapon available in the form of spores that can be put into powders, sprays, water or food and easily distributed in the environment. In the Amerithrax attacks of 2001, as they became known, 22 people in the US were infected by the substance and five of them died. Joseph Persichini, Jr., Acting Assistant Director in Charge of the FBI, has said that

> [T]he scientific advances gained from this investigation
> are unprecedented and have greatly strengthened the

government's ability to prepare for—and prevent—biological attacks in the future. Since the first anthrax mailing, investigators have worked hand in hand with the scientific community to both solve this case, prevent another and to be best positioned should another occur.

In our own corner of the world, the events at the school and the airport presented a huge learning exercise for the Defence Forces, and in particular the Ordnance Corps, but also for AGS, first responders and medics. We were beginning to develop a capability in the area of CBRN defence, but it was very early days. For example, at the school callout, the fire brigade had decontamination showers on their trucks, but we had no idea how to assemble them; our testing equipment wasn't up to date, and the structures put in place to deal with standard callouts weren't fit for purpose. Disposing of a bomb requires a team of three: the bomb-disposal officer, their deputy and a driver. In CBRN, a team of about seven is needed, with backups. Also, the equipment you need is quite different, such as the suits, which aren't the bulky bomb-disposal suits, but much more lightweight – but which need to offer protection from CBRN agents. Another innovation is the use of motorised transport such as a Segway at the scene: it can look daft, but when you're dealing with a device that could be CBRN, time is of the essence. With oxygen supplies limited, an operator doesn't want to waste time walking back and forward.

Of course, with CBRN weapons, you want the minimum number of personnel exposed, which further increases pressure on the operators. In fact, a number of years after the school incident, there was another incident involving anthrax, this time in Limerick Prison. A convicted criminal, Essam Eid, known for his involvement in the 'Lying Eyes' case involving the attempted murders of three men, had smuggled anthrax into the prison in his contact-lens case. The CBRN team set to work, expecting a negative result, but when they checked with their updated equipment, the substance turned out to be the real thing.

Quite apart from the risk of CBRN attacks, in our increasingly fragmented and unpredictable world, another new threat has emerged, that of the 'marauding terrorist'. A marauding terrorist is someone who is set loose with a weapon or bomb (or both) who is ready to inflict maximum damage and to die in the process. Unfortunately, we have seen many examples of this, from the attacks in Mumbai in 2008, when a number of hotels and train stations were attacked and 166 people died, to the *Charlie Hebdo* attacks in January 2015 and the later dreadful attacks in Paris that year. Like many US mass-shooter events, these attacks offer the potential for continuing casualties until the perpetrator(s) are physically stopped. Nobody wants to dwell on terrible events like the massacre at the Bataclan, but if we are to stand any chance of tackling this new and complex problem then we need to examine and learn from them.

When I was in Afghanistan, a colleague lent me a US military publication entitled *The Insider Threat* – which taught me a lot. The situations given were, for example, when an Afghan would turn on a US soldier. (Nowadays, insider threats can come in the form of hacking or the installation of malware on computers, either by employees or 'bad actors', which isn't really in my area of expertise, but probably presents one of the biggest security threats today.) In Afghanistan, the US-produced document advocated the use of what was called Advanced Situational Awareness Training (ASAT) – or Human Behaviour Pattern Analysis. Basically, this means using our own powers of observation in a number of areas to identify people who might pose a threat. Let me give you an example. In the 2016 attacks in Brussels Airport, the taxi driver who brought the bombers to the airport noticed that they were sweating profusely when they got into the car. Secondly, there was such a powerful chemical smell in the car that the driver thought he might pass out. And thirdly, when it came time to unload their bags, one of them was so heavy the strap of the bag tore. Taken together all of these elements were major red flags, and you might wonder why he didn't pick up on them at the time, but the answer probably is that he wasn't looking for them. According to the BBC, in the aftermath of the attacks, police were approached by that same taxi driver who said he had driven three men with big bags to the airport on the morning of the attacks. Local media reported that the driver had refused to take one of the men's large bags because there was not enough room in the vehicle.

The purpose of this example is not to blame someone, but simply to say that ASAT requires an ability to recognise when something doesn't seem right and to act on it in the moment.

A similar situation occurred at the Manchester Arena, when a suicide bomber killed 22 people at an Ariana Grande concert in 2017. As Sky News later reported, the public inquiry found that there were 'unacceptable and unjustified' security failures that night. Two police officers left a room unsupervised, where the bomber was able to hide for over an hour. In fact, they'd walked by the bomber outside the building, a large backpack on his back, to go on a lunchbreak. Now, plenty of people in the area were wearing backpacks – the Arena is beside Victoria Station – but one police officer later reflected that there was something about this particular backpack that should have attracted her attention. Furthermore, a security guard from a private company noticed him as well, but didn't follow his instincts to investigate further, telling the inquiry that he had a 'bad feeling' about the bomber, but 'I was scared of being wrong and being branded a racist if I got it wrong and would have got into trouble'. The issue here is not one of racial profiling: security experts told the inquiry that the security guards at the venue 'should have been specifically and clearly told in briefings what to do if a member of the public informed them about suspicious behaviour', but 'it was not clear that that happened'. Again it's not about pointing the finger at people and blaming them: it's about learning lessons from terrorist events like this to make sure that they don't happen again.

At the end of *The Insider Threat*, in the references, I came across the name of Greg Williams and his expertise in Human Behaviour Pattern Analysis, and this intrigued me: I feel that everyone should be introduced to the concept of identifying potential threats and assessing whether they are going to lead anywhere or not. I often hear the criticism that we are making people paranoid, but if those security guards had recognised the threat that the suicide bomber posed in Manchester, could lives have been saved? In all likelihood, yes.

One lighter example of ASAT came when I had the opportunity to test the then-Secretary-General of the UN, Ban Ki-moon, on his situational awareness. He was to visit the UN school in the Curragh in 2018. Now, I was getting on at this stage and a bit jaded, so I called in my captain to my office in the Ordnance School and said, 'You're giving a demo to the Secretary-General of the UN in the Curragh. Go down, recce the place, talk to them and tell me what you plan to do.' Off he went and I got back to whatever it was I was doing.

He came back an hour later. 'Sir, I've bad news for you.'

'What?'

'The Chief of Staff himself has decreed that you're giving the demonstration. After all, it is the Secretary-General of the UN.'

'Fine,' I sighed. 'I'd better obey, then.'

At the UN training centre at the Curragh, there's an observation post built to the same specifications you would see in places like the Lebanon in order to get people accustomed to life in those buildings. I went and had a look at it and said, 'Okay, we're going to do a demonstration of our robots and then we'll have some IEDs and put them there like that … And then I'm going to give him a briefing about the marauding-terrorist course and talk about situational awareness. Then I'm going to bury an IED in the ground under the UN position and I'm going to remove it using a robot.' We rehearsed it and then I went away, and it was all sorted.

In the morning, we went down to the Curragh and there he was, with all of the Defence Force chiefs of staff and then-Minister for Defence Simon Coveney. The Secretary-General did the usual inspections and planted a tree, then he arrived in front of me. I introduced myself and my team and started going through all of the equipment, demonstrating various mock IEDs and our robots in action. He listened politely, but when I got to marauding terrorism, he started to look interested. 'What do you teach in the course?' he asked.

'We teach people the ability to observe what's happening around them. You don't need equipment, robots or anything for this. Instead, we train the brain so that it becomes situationally aware.' I had used the catchphrase 'absence of the normal, presence of the abnormal', which I will have inscribed on my gravestone!

He wasn't quite getting it, because it was all a bit abstract, so I said, 'May I give you an example? This morning, whatever hotel you were staying in, your wife gets out of bed and goes and gets her hair done. She comes back and you don't even notice it. That's a lack of situational awareness.' Now, while this was going on, Ban Ki-moon's wife walked up behind him, and she tapped him on the shoulder. 'That's exactly what happened this morning!' He looked at me and said, 'I get it.'

———◆———

With fresh enthusiasm, and with great support from Colonel Brian Dowling, Colonel David Sexton and Peter Daly, we at the Ordnance School decided to run a course on marauding-terrorist events, to educate scene commanders. It was superb, pulling together the knowledge of organisations like the FBI, the ATF, the Royal Canadian Mounted Police, the Los Angeles Sheriff's Department, the Metropolitan Police and ASAT trainers to produce a range of options for dealing with this ever-evolving threat. The key is 'a range of options': with a host of different threats coming from many different sources, the solution can't be one-size-fits-all. In my presentation on the first course, I said, '[T]his course ... offers no singular system to be adhered to, copied and utilised; it is a gathering of open minds, working with no fear, to discuss, argue and learn, to be bold, disruptive and innovative with no agenda ... working to save lives by enhancing the approach to marauding terrorism and future-proofing it.'

Lofty words, but the reality taught us a lot. Together with my colleague Captain Alan Kearney, we organised a simulated attack in the Curragh, which was to act as a village into which a marauding terrorist would drive at speed and kill people with their vehicle. We hired people to play the roles of ordinary civilians, we bought prams and dolls, shopping bags and so on. The exercise was timed with a huge clock, with the minutes running down as the Gardaí came in on helicopters and our 'terrorists' pretended to shoot. The initial AGS response was unarmed, which left them unable to render aid to the casualties, and they had to leave the scene. This produced nervous laughter from the VIP attendees. The Garda Armed Support Unit followed, but with two officers, their response was limited. Finally, the Garda ERU (Emergency Response Unit) in Dublin were tasked and arrived by military helicopter. They engaged the terrorists. However, at this stage there were many casualties and the terrorists had locked themselves in a building which was now on fire.

To explain why the response came solely from the Gardaí, the army isn't able to respond rapidly to an unfolding terrorist event. Units like Special Forces, for example, require a lead-in time due to the capability they bring. EOD can respond rapidly but wasn't, at that time, prepared for dynamic scenarios with terrorists still active at the scene. This came as a surprise to many: that the police, and not the army, had had to engage in a mini-battle against well-prepared terrorists.

At the end of the exercise, there were large numbers of casualties. Had this been a real-life situation, it would have been a disaster, but it provided us with a huge amount of information. We went into the classroom then and we went through it, working out what had and hadn't worked. With our military culture, our response is slow and measured and unsuited to countering an evolving threat. The police were much faster, but their two teams didn't integrate. As it has been put, 'The army trains and plans to fight, and does very little; the police spend all their time fighting, with little time training or planning.'

We also learned that there was no dynamic risk assessment – i.e. working out what was happening as it was happening, and making decisions accordingly. Some of our medical equipment wasn't adequate to stop people bleeding to death (again, this was simulated), and we realised that we needed simple kits that ordinary people could use to help the injured before the emergency service response. And then we went again, this time in Dundrum Shopping Centre and in Dublin Airport, testing out our response again.

With the help of UK operators, Alan had trained law-enforcement officers in the USA for a number of years, integrating counter-IED (C-IED) into their Active Shooter programme. He learned a great deal from the experience, and what he learned formed the basis of the Ordnance School training programme. Alan harnessed our exceptional C-IED network to move the initiative to the EU and NATO stage. Like C-IED before it, it was more than just an offering of training courses: it was a drive for excellence to meet a threat that was emerging. A threat that many could not, and still do not, see.

In Kearney's report on the first course in 2017, he noted the classic response to armed events centres around containment, negotiation and/or tactical engagement and resolution. Once an emergency involving firearms and/or explosives is in progress, it is left to armed police, and ideally police tactical units, to deal with the antagonists. Meanwhile, the remaining emergency service providers 'sit it out' until they receive the all-clear. He pointed out the clear flaws in this approach. For example, in the US, school shooters had been left to roam buildings killing while the security services waited outside. Kearney identified the UK-based MTFA (Marauding Terrorist Firearms Attack) programme as a framework for development, along with aspects of the C-IED and Active Shooter programmes. 'The objective of the framework is to remove many key decisions from the scene commander, decisions which have, in the past, cost lives.'

The other thing we needed to develop was the psychological side of things. Here, forensic psychologist Kiran Sarma from the University of Galway helped us to understand how best to manage our emotions in emergency situations, so that they didn't cloud judgement. This applies to ordinary citizens as well as professionals, of course, but in our course, we were looking at responders who would have to make key decisions in a rapidly unfolding situation.

As Sarma told us, 'The managing of emotions during a terrorist attack involves a battle between uncontrolled cognitive and emotional processes on the one hand and controlled practised thinking on the other.' This was labelled 'thinking fast' and 'thinking slow'. Now, you might reason that quick thinking would save the day, but no: quick thinking involves us falling back on old patterns of behaviour that might not help us in a new situation – for example, freezing in the face of something unexpected. 'Slow' thinking allows us to settle our emotions before making a decision, and 'is associated with having a greater belief in one's ability to address the problem, to manage difficult emotions and thus to identify fruitful actions'.

The course was a great success and it made me realise that Ireland really does have something to offer in key areas. We don't have the resources to focus on conventional warfare – i.e. tanks, infantry and so on – but we are very good at specialist ops: remember the Army Rangers and Navy interception of the drugs vessel in Cork in September 2023? We also have great expertise in counter-IED. So, I decided to run two prototype bomb-disposal courses, based on my experience in Afghanistan, expanding our expertise from simple bomb disposal to broader counter-IED.

Again, we needed to develop some scenarios to learn from, so we turned Kerry into a hostile area and designed a state of insurgency there. I stress that this was an exercise! The night before, in a tent on the Cork/Kerry border, the commander of the exercise had to integrate all the different assets into a coordinated plan to get a hostage released. We rehearsed potential scenarios ad nauseam, so that everybody knew their jobs, and off we went. Our scenario was that a hostage had been taken and held in a remote location. A video was sent demanding a ransom, and the room in which he was held was booby trapped. So, we needed to gather intelligence with our drones to confirm that he was being held there – without attracting the kidnappers' attention. If the drone fell out of the sky and landed on the house, they'd know we were onto them.

We also staged exercises in Dublin Airport, Dundrum Shopping Centre and Heuston Station, and we got great support from everyone there. We tested lots of possible scenarios, for example, a possible device in one of the restaurants in Dublin Airport, which presented us with all kinds of interesting data. One of our course attendees wanted the airport closed down to deal with the incident. The airport police told him, 'That's impossible. We can't close down an international airport, but we could isolate part of it.' Then he had to gauge how much of the airport to clear, because of the size of the bag and what it might contain in terms of explosive. There's a formula, and then he had to go through the options of remote, semi-remote and manual. This whole process took a number of hours, while 'passengers' and airport staff were complaining about closed runways and delays. I reminded him of the one great rule of bomb disposal: if you are in charge, you isolate yourself, so that you can make decisions with a clear head. That's what your deputy is for: to deal with all of the hassle while you work out a plan.

Another area in which Ireland used to lead, thanks to its experience with the IRA, was that of homemade explosives. If you remember, when Colonel Gaddafi was in charge in Libya, he supplied the organisation with Semtex, but when this was no longer available, the organisation had to come up with new ideas.

So, ammonium nitrate in fertiliser became their go-to, mixed with different substances to create different effects. For example, mixed with icing sugar, it can cause a huge flash burn, whereas ammonium nitrate mixed with fuel oil is smelly, dirty and pretty horrible, but effective. You might ask, why not reduce the amount of nitrogen in fertiliser? That's what manufacturers did, but the IRA worked out how to recrystallise it and make it powerful again. Now, peroxide-based explosives are the focus. All of the terrorist attacks of the 2000s used TATP, or triacetone triperoxide. It's popular because its base, peroxide, is readily available. It's very dangerous and not easy to make, but it is doable, and the results have been catastrophic.

In recent years, the FBI has taken the lead in this area, because they have a fantastic training school and brilliant scientists. So, in 2014, we decided to see if we could use their expertise to get back in the game. In two innovative courses, we brought military and civilian organisations, such as the Armed Forces, Europol, the FBI, the ATF and the EDA (European Defence Agency), together to look at the area of homemade explosives. We had more than 30 students from 16 countries, and we brought in literally everything that would be needed to make explosives. We set up a clandestine lab, with cookers, ovens and chemicals, and we sent people in to work out whether they were manufacturing explosives there or drugs. We spent a lot of time learning about TATP, how to make it, what chemical process is involved; we dismantled it, dissolved it, moved it, learned everything there was to learn about this deadly new explosive.

What did the course teach me? How, with proper vision and defined outcomes, you could integrate military and civilian procedures to be effective. This template could be used to counter IED and CBRN weapons. To this end, 'no stove piping' became the motto. That meant that information should not be shared without the proper context. It was to be shared openly and lessons learned without affecting anyone's area of responsibilities. It just so happened that after the course, we sent a team out on a callout to what they'd been told was an explosives lab. The officer in charge on the day had been on the course and quickly worked out that it was a drugs lab and that the correct people to take charge were the Gardaí.

On my retirement in May 2018, I was asked to brief the senior command and staff on my reflections past and present.

By the time I retired, I had amassed a total of 45 years in the Defence Forces. Writing that down now, that feels like a very long time! As I looked back, I came to understand that in my mind, I had had 35 years of learning at home and abroad and 10 years in the Ordnance School to apply the lessons and see where they could take us. Now, my job was done.

I bore the physical and mental scars of my service. Like so many of my colleagues in bomb disposal, I have tinnitus, sometimes disabling vertigo and a condition known as MTBI, or mild traumatic brain injury. I hadn't heard of this until I was diagnosed in 2021, after spending a month in bed due to my problems with balance. I'm not complaining, simply stating that this and the mental strain of often stressful and difficult situations has taken its toll. Quite honestly, I would do it all again. What I gained in terms of camaraderie, team spirit and support has far outweighed the costs.

However, when briefing my colleagues in 2018, one word stuck in my mind. Relevance. We need to make the Defence Forces relevant to the needs of a modern Ireland while maintaining our neutrality; we also need to resource the organisation to meet specific global security threats. Sadly, while we have made some progress, we have a long way to go.

It was, and is, my view that the Defence Forces can play a vital role in the community. I drove through the Curragh recently and it saddened me to see how much it has declined since the formation of the State, when 6,000 British soldiers were stationed there and 4,000 civilians – and, indeed, since my time, when it was basically a town built to serve the camp. When I was a cadet, in 1973, the place had a cinema, shops, even its own abattoir. Now, the officers' mess is boarded up and the shops and cinema are closed. Our navy has six ships, but only half of them can function because we can't recruit people to join the navy. And if we do, the private sector can lure trained engineers away with the promise of a greater salary.

The above is symbolic of the state of the Defence Forces presently. And yet we can work with academics and industry to develop new and innovative responses to the many threats that present themselves now. For example, in 2012, we approached University of Galway Professor of Glycoscience, Lokesh Joshi, to help us to develop PCT, or 'pathogen capturing technology'. The resulting all-natural substance, based on milk, allows for pathogens to be 'Velcroed' up via wipes and a spray, so that they can be removed completely, but – and this is crucial – the actual substance remains, so that it can be analysed. However, what was intended to be a substance capable of removing biohazards such as anthrax has also been discovered to have much wider applications, and companies in the wound care, skincare and cosmetics industry have come on board to see if it can be used in other ways. One example of using key areas of expertise to play our part on the world stage.

We are a small country, so we don't need a huge military force bristling with equipment, but we do need a properly resourced, agile and well-paid one, so that we can continue to play our part in future peacekeeping missions, in counter-IED/specialist search and special force operations, and in NATO – which we can do without compromising our neutrality. It's not rocket science. Ireland is the second richest country in the EU and yet we spend 0.26 % of our GDP (that's €1.25 billion, including pay and pensions) on defence. This is the lowest proportion in Europe – something we should not be proud of as a country. Nor should we be proud of the fact that as servants of the State, our strength has diminished from 13,000 in the 1970s to 7,500 in May 2024. Historically our Defence Forces have been at the lowest when needed most – in 1939, 1969 and 2024.

It is my view that our overseas missions have become overly UN-centric, which is not where a modern military should be positioned. The country's alignment with neutrality is a key cornerstone of our defence and security policy, which I support wholeheartedly, but many people, including myself, have served alongside NATO, for example in Afghanistan, without compromising our neutrality. I have served with the UN, the EU and with NATO and understand that the tricolour on a soldier's shoulder is such a valuable asset it should not be diluted. However, we can still serve in theatres of conflict using specific skillsets, which benefits us and the organisations with which we serve.

One would have expected that with the instability in Europe and globally, the government would analyse these threats and plan effective responses. For example, can we protect subsea energy and communications critical underwater infrastructure enablers? Can we protect our skies? To do this, we need a national security strategy, and yet although work began on this in 2019, no strategy has yet emerged.

TD Cathal Berry, a former Army Ranger, was clear in a 2023 interview with Niall O'Connor in the *Journal*: 'The fact that there is no strategy document is part of a major cultural problem – it is the biggest security problem here – they simply don't take it seriously … there are multiple geopolitical risks at the moment, West Africa, Ukraine, the behaviour of countries like Russia and China – all of that is happening but yet all Ireland does is pay lip service to national security.' I agree. There has been a failure to understand and to take seriously the very volatile security landscape in which we exist.

In a 2022 article for *Foreign Policy*, Eoin Drea put it more starkly: 'Ireland is Europe's weakest link … don't count on Ireland to have any part in countering or deterring real and potential Russian threats. With a token navy of six active patrol vessels, not a single submarine to cover its vast marine zone, and annual defense [*sic*] spending of barely more than 1 billion euros (about 0.3 percent of GDP), Ireland stands out as the worst-prepared European country to meet any meaningful threat – or even anything less than a meaningful threat.' Drea went on to quote from our own committee into the Defence Forces: 'The absence or near-absence of crucial capabilities – including intelligence, cyberdefense [*sic*], radar, intercept jets, and heavy airlift planes – render the Irish forces "unable to conduct a meaningful defence of the State against a sustained act of aggression from a conventional military force."' Furthermore, 'With barely 700 staff, the Irish Air Force (known as the Aer Corps) lacks the capability even to track aircraft across Irish skies. Irish parliamentarian Tom Clonan, a former army officer, noted that Ireland is "the only country in the EU that cannot monitor its own airspace by primary radar. Nor can Ireland patrol its own airspace with even the most basic jet interceptor."'

This makes for alarming reading, doesn't it? The fact that the Irish State is considered a 'freerider' or 'freeloader' should be a source of national shame, the result of a quarter-century of whole-of-government neglect of effective and relevant Defence Forces capability development. At the same time we have seen the sad degradation of military infrastructure and property management: my trip through the Curragh has shown me this. And yet, many reviews have focused on reorganising the Defence Forces, which to me does not address the problem. All it does is increase the frustration of all serving and retired personnel.

There are many decisions over the last number of years which have highlighted the lack of defence-force strategic focus. Two decisions in particular stood out to me. The decision to implement a 'promotion on merit' system for the rank of commandant and above has, ironically, neutralised many officers' real potential. Quite apart from the fact that it's difficult to assess 'merit' in peacetime, arguably some are afraid to stick their necks out and join an operational unit lest something go wrong and jeopardise promotion. I would argue that, up to the rank of captain, we have a cohort of excellent officers. Above this rank, the focus and energy are devoted to 'the next step' rather than on developing real leadership skills.

The second poor decision came with the closure of the Defence Forces' apprentice school in Naas. This was an outstanding educational institution recognised by industry and academia as producing soldiers of exceptional talent and expertise, but also producing people who could make a real and valuable contribution to Irish society. I can't understand the logic of this decision.

The Irish Defence Forces have been subject to many reorganisations and reviews, and yet the Department of Defence remains static. There are many helpful and supportive people working in the Department; however, the person who holds the purse strings has the power and control, and that currently rests with the secretary general in the Department, and not the Defence Forces' Chief of Staff – unlike in, say, An Garda Síochána. While I agree with the focus on the Defence Forces in pursuing reform, this should be matched with equal focus on the Department of Defence.

The Irish Defence Force is a unique organisation with some exceptional people in it, with many and varied skillsets. It has so much to offer Ireland, Europe and the world. It could be a high-value organisation, which with effective leadership would be a significant asset. However, we need to remove the negative process and the risk-averse culture in the organisation and to develop positive processes and dynamic structures which will assist in making both the Defence Forces and the Department relevant.

At an operational and tactical level we have shown what one small part of the Defence Forces – the Ordnance Corps – has achieved since 2010: a total of 33 countries represented on NATO/EU/UN/EDA courses in Ireland. The Ordnance School was also chair of an international working group on C-IED, supported by China. What's more, we have two indigenous robot companies in Ireland in ICP Newtech and Reamda, producing some of the best humanitarian unmanned ground vehicles in the world, employing Irish people. Both Canada and Germany have supplied Irish robots for humanitarian demining in Ukraine. In fact, Germany's purchase was funded by EU funds. I am hoping the Irish government, through the department of foreign affairs, will follow suit.

I am an evaluator on the European Defence Fund, to which Ireland has contributed €150m over a six-year period. However, we aren't serious players in the fund, and we should be. As an evaluator, I see the benefit of actively supporting the fund. It brings industry/academia and the military together to develop capability covering most security strands. It will create jobs and increase Ireland's security profile. It does not in any way infringe on neutrality. To this end, we need a small, dynamic and qualified group of people to support this work. We need to normalise our discussion on defence and have real communication with the public. I think that as a country we are afraid to discuss these issues openly. As I said before, I do not want Ireland to join NATO – and there's no need for us to do so – but I do want the Defence Forces to have the capability and resources available to support the global security agenda in a dynamic manner with committed, effective leadership.

The recent decision by the government to join two Permanent Structured Cooperation projects (PESCO) and two EDA projects looking at Critical Seabed Infrastructure, environmental management and cyber is a welcome development. However, it is time to send qualified Defence Forces personnel to the EDA on a permanent basis, to maximise its potential for Ireland and ensure that the results and capability developed are relevant and timely.

We have so much potential in Ireland in the domain of security and defence: we need to develop it and to take our role on the world stage with courage and commitment. When we do this, we can achieve great things.

ACKNOWLEDGEMENTS

When I met Teresa Daly for the first time to discuss the possibility of writing a book, I really needed convincing. My story is really no different from the stories of the many wonderful men and women who have soldiered in similar areas in the Defence Forces. However, once Teresa introduced me to Alison Walsh, I knew I was in safe and talented hands. Alison's ability to get to the core of the different areas we discussed was so impressive. She had no prior knowledge of the Defence Forces and its unique nature and culture (although she does share a fence with Cathal Brugha Barracks!). Despite this, it wasn't long until she was using military terms in our discussions in a manner that would indicate that she was using them all her life.

I would like to thank all the wonderful staff at Gill, including Teresa Daly, Fiona Murphy, Margaret Farrelly and Iollann Ó Murchú, for their absolute professionalism and support. All our work together was carried out in a positive and friendly manner. Gill is fortunate to have people of such quality.

I would like to thank the close friends and family who assisted at various times in stimulating the 'grey matter'.

My family have made sacrifices with my absence over the years, as have most of the families of members of the Defence Forces. In many cases this goes unnoticed until something tragically goes wrong.

The Defence Forces is a unique institution, flawed (just like society) but with the most amazing people, with amazing qualities and skill sets. It's a pity that politicians and senior leadership continue to fail to see their strengths and offer real support. They are at times a little too hasty in attacking their weaknesses.

The Defence Forces should be trained and equipped to meet the demands of today and to become future-proofed for the emerging threats of tomorrow. It is, potentially, an incalculable asset to Ireland.

GLOSSARY OF TERMS

ASYMMETRIC/HYBRID THREATS – use of unconventional methods to negate an opponent's strengths while exploiting their weaknesses

ANSF – Afghan National Security Forces

AO – area of operations

ASAT – Advanced Situational Awareness Training

ATO – Ammunition Technical Officer

ATF – Bureau of Alcohol, Tobacco, Firearms and Explosives (US agency)

CBRN – chemical biological radiological nuclear

CBRNIDD – chemical biological radiological nuclear improvised device disposal

C-IED – counter improvised explosive device

CO – Commanding Officer

COO – Chief Operating Officer

CUA – Commander's Update Assessment

CUI – critical underwater infrastructure

DFI (ÓGLAIGH NA H-ÉIREANN) – Defence Forces Ireland

ECMM – European Commission Monitoring Mission (renamed the European Union Monitoring Mission (EUMM) in 2000)

EDA – European Defence Agency

EOD – explosive ordnance disposal

FCA – Fórsa Cosanta Áitiúil (name of the land component of the army reserve forces prior to 2005)

HOBO ROBOT – an unmanned ground vehicle used in bomb disposal, so-called because it wanders. A similar bomb-disposal robot was given the name 'Vagabond'.

HVO – former main military force of the Croats of Bosnia and Herzegovina.

IDF – Israel Defense Forces

ICP – incident control point

ICTY – International Criminal Tribunal for the former Yugoslavia

IED – improvised explosive device

IEDD – improvised explosive device disposal

INCOMING, OUTGOING – shells coming in and shells going out

ISAF – International Security Assistance Force (Afghanistan)

JIEDDO – Joint Improvised Explosive Device Disposal Organisation (USA)

NATO – North Atlantic Treaty Organization

NDS – National Directorate of Security (the primary intelligence organ of Afghanistan, now dissolved)

OSCE – Organisation of Security Cooperation in Europe

RICKSHAW – radio sign for ordnance officer

RSP – render-safe procedure

TATP – triacetone triperoxide (chemical used in explosives)

TPU – time and power unit (one of the components of a bomb)

UAV – unmanned air vehicle

UGV – unmanned ground vehicle

UN – United Nations

UNDOF – United Nations Disengagement Observer Force

UNIFIL – United Nations Interim Force In Lebanon

UNOPS – United Nations Office for Project Services

UVIED – under-vehicle improvised explosive device

WILCO – will comply (radio procedure word)